CONVEXITY
AND RELATED
COMBINATORIAL GEOMETRY

PURE AND APPLIED MATHEMATICS

A Program of Monographs, Textbooks, and Lecture Notes

Contributions to *Lecture Notes in Pure and Applied Mathematics* are reproduced by direct photography of the author's typewritten manuscript. Potential authors are advised to submit preliminary manuscripts for review purposes. After acceptance, the author is responsible for preparing the final manuscript in camera-ready form, suitable for direct reproduction. Marcel Dekker, Inc. will furnish instructions to authors and special typing paper. Sample pages are reviewed and returned with our suggestions to assure quality control and the most attractive rendering of your manuscript. The publisher will also be happy to supervise and assist in all stages of the preparation of your camera-ready manuscript.

LECTURE NOTES
IN PURE AND APPLIED MATHEMATICS

Other Volumes in Preparation

CONVEXITY AND RELATED COMBINATORIAL GEOMETRY

Proceedings of the
Second University
of Oklahoma Conference

edited by
David C. Kay and Marilyn Breen
Department of Mathematics
University of Oklahoma
Norman, Oklahoma

MARCEL DEKKER, INC.   New York and Basel

Library of Congress Cataloging in Publication Data
Main entry under title:

Convexity and related combinatorial geometry.

   (Lecture notes in pure and applied mathematics ;
v. 76)
   Includes index.
   1. Convex polyhedra--Congresses. 2. Combinatorial
geometry--Congresses. I. Kay, David C., [date].
II. Breen, Marilyn, [date]. III. Series.
QA640.3.C66   516.3'6   82-1381
ISBN 0-8247-1278-1   AACR2

MARCEL DEKKER, INC.
270 Madison Avenue, New York, New York 10016

Current printing (last digit):
10 9 8 7 6 5 4 3 2 1

PRINTED IN THE UNITED STATES OF AMERICA

PREFACE

The contents of this work consist of the papers presented at a convexity conference held at the University of Oklahoma, March 13-15, 1980. The participants were particularly fortunate to hear the presentations of results and surveys by the keynote speakers Victor Klee, David Barnette, Don Chakerian, Tom Sallee, and John Reay (whose papers appear in these proceedings). All the papers presented were worthwhile and interesting, but particularly noteworthy were (1) the outstanding survey of the d-step conjecture presented by Victor Klee (approached from an ad hoc point of view to promote the breaking away from traditional and unfruitful attacks on the problem), (2) the solution by Carl Lee of the longstanding characterization problem of the family of f-vectors of simplicial polytopes in R^d via the Dehn-Sommerville equations (one half of the characterization--the necessity of the conditions--was accomplished earlier by R. Stanley, referenced in Lee's work), and (3) a very useful organization of the multitude of scattered results and ideas of abstract convexity and some new relationships in that field by Robert Jamison-Waldner.

All the papers appearing here have been accepted prima-facie without the aid of referees, so some authors may have published these results in a different form in referreed journals.

In the opinion of its participants, the conference was an important and delightful success, highlighted by a telegram of good wishes from Germany by colleagues in the field, Ludwig Danzer, J. Eckoff, U. Wegner and C. Zamfirescu.

Gratitude is sent to our benefactors, without whose support the conference could not have materialized: The J. Clarence Karcher Foundation at the University of Oklahoma which provided the major funding for the conference, and the National Science Foundation which provided extra participant support. Thanks go also to the expert job by the typists, Trish Abolins and Detia Roe.

David C. Kay
Marilyn Breen

CONTENTS

CONTRIBUTORS

David Barnette, *Department of Mathematics, University of California at Davis, Davis, California.*

Arne Brøndsted, *Institute of Mathematics, University of Copenhagen, Copenhagen, Denmark.*

G. D. Chakerian, *Department of Mathematics, University of California at Davis, Davis, California.*

René Fourneau, *Unité de Mathématiques, Institut Supérieur Industriel Liégeois, Liège, Belgium.*

Paul Goodey,[*] *Department of Mathematics, University of Oklahoma, Norman, Oklahoma.*

Jacob E. Goodman, *Department of Mathematics, The City College, City University of New York, New York, New York.*

Hans Herda,[†] *Department of Applied Mathematics, the Weizmann Institute of Science, Rehovat, Israel.*

Robert E. Jamison-Waldner, *Department of Mathematical Sciences, Clemson University, Clemson, South Carolina.*

Victor Klee, *Department of Mathematics, University of Washington, Seattle, Washington.*

Jim Lawrence, *Department of Mathematics, University of Kentucky, Lexington, Kentucky.*

Carl Lee,[‡] *Center for Applied Mathematics, Cornell University, Ithaca, New York.*

[*]Current affiliation: *Department of Mathematics, Royal Holloway College, London University, London, England.*

[†]Current affiliation: *Department of Mathematics, Boston State College, Boston, Massachusetts.*

[‡]Current affiliation: *Department of Mathematical Sciences, IBM T. J. Watson Research Center, Yorktown Heights, New York; and Department of Mathematical Sciences, University of Kentucky, Lexington, Kentucky.*

Erwin Lutwak, *Department of Mathematics, Polytechnic Institute of New York, Brooklyn, New York.*

Joseph Malkevitch, *Department of Mathematics, York College, Jamaica, New York.*

Bruce B. Peterson, *Department of Mathematics, Middlebury College, Middlebury, Vermont.*

Richard Pollack, *Department of Mathematics, Courant Institute of Mathematics, New York University, New York, New York.*

John R. Reay, *Department of Mathematics and Computer Science, Western Washington University, Bellingham, Washington.*

G. Thomas Sallee, *Department of Mathematics, University of California at Davis, Davis, California.*

Gerald Sierksma, *Subdepartment of Mathematics, Econometric Institute, University of Groningen, Groningen, The Netherlands.*

Andrew Sobczyk, *Department of Mathematical Sciences, Clemson University, Clemson, South Carolina.*

Wolfgang Spiegel, *Fachbereich 7 (Mathematik), Gesamthochschule Wuppertal, Wuppertal, Federal Republic of Germany.*

Philip H. Turner,[*] *Department of Mathematics, Louisiana State University, Baton Rouge, Louisiana.*

[*]Current affiliation: *Department of Systems Forecast, United Illuminating Company, New Haven, Connecticut.*

HOW MANY STEPS?

Victor Klee

Department of Mathematics
University of Washington
Seattle, Washington

Stimulated since the early 1950's by its relationship to linear programming
and more recently by other connections with computational questions, the
combinatorial study of convex polytopes has advanced greatly in the last
25 years. For an overview the reader may skim successively the survey articles or books of Klee 1966 [14], Grünbaum 1967 [8], Grünbaum and Shephard
1969 [11], Grünbaum 1970 [9], McMullen and Shephard 1971 [21], Grünbaum 1975
[10], Klee 1975 [15], and Barnette 1978 [3] — and then read the recent
proofs by Billera and Lee [4] and Stanley [23] of an important 1971 conjecture of McMullen [20].

The d-step conjecture was first formulated by W. M. Hirsch in 1957 (see
Dantzig [5,6]), and despite progress on many other fronts it remains unsettled. Because of intrinsic interest and connections with questions of computational complexity, and because solutions may well require the development
of deep new methods, the d-step conjecture and its relatives are probably the
most important open problems on the combinatorial structure of convex polytopes. The present note, which is extracted from a longer article in preparation, states several equivalent forms of the d-step conjecture. Some deal
explicitly with edge-paths on polytopes, one involves matrix pivot operations,
and one concerns an exchange process for simplicial bases.

1

Suppose that X is the nonnegative orthant R_+^d in real d-space R^d, x is the point $(0,0,\ldots,0)$, x' is the point $(1,1,\ldots,1)$, and $X' = x' - R_+^d$, a translate of the nonpositive orthant $-R_+^d$. Then

(a)   X and x' are affine orthants with respective vertices $x \in$ int X' and
      $x' \in$ int X , and

(b)   the intersection $P = X \cap X'$ is bounded.

One form of the d-step conjecture asserts that *whenever conditions* (a) *and* (b) *are satisfied the vertices* x *and* x' *of the polytope* P *can be joined by a path formed from at most* d *edges of* P. That is certainly true in the example, for there P is the unit cube $[0,1]^d$. However, even when (a) and (b) are augmented by the requirement that P is simple (each vertex incident to precisely d edges) there are many other possibilities for P. In particular, the possible number of vertices then ranges from $d^2 - d + 2$ to $4\binom{3k-1}{k}$ when $d = 2k$ and to $2\binom{3k+1}{k}$ when $d = 2k + 1$.

The d-polytopes of the above form $X \cap X'$ have precisely 2d facets (faces of dimension d - 1). Hence a formal strengthening of the above conjecture is the conjecture that $\Delta(d,2d) = d$, where $\Delta(d,n)$ denotes the maximum diameter of d-polytopes P with n facets. (That is, $\Delta(d,n)$ is the smallest integer k such that any two vertices x and x' of such a P can be joined by a path formed from at most k edges of P.) A further formal strengthening is the conjecture that $\Delta(d,n) \le n - d$. However, it was proved by Klee and Walkup [17] that these conjectures are all equivalent, though not necessarily on a dimension-for-dimension basis. Another equivalent conjecture is that *any two vertices of a simple polytope can be joined by a path that does not revisit any facet* [17].

Turning now to matrix pivot operations, we note that the d-step conjecture is equivalent to the following:
If the real d × (2d + 1) *matrices* A = (I,B,c) *and* A' = (B',I,c') *are row-equivalent, where* $d \ge 2$, I *is the* d × d *identity matrix, and the columns* c *and* c' *are* > 0 , *and if the polyhedron*

$$P = \{x \in R_+^{2d} : (I,B)x = c\}$$

is bounded, then it is possible to pass from A *to* A' *by a sequence of* $\le$ d *feasible pivots followed if necessary by a permutation of rows.*

Here a *pivot*, as applied to an $m \times (n + 1)$ matrix $S = (s_{ij})$, is the operation
of choosing (i,j) with $j \leq n$ and $s_{ij} \neq 0$, then dividing the i-th row of S by
s_{ij} so as to obtain 1 in position (i,j), and finally subtracting appropriate multiples of the i-th row from other rows so as to obtain 0 in all
positions (h,j) for $h \neq i$. A pivot is *feasible* if the last column of the
matrix is nonnegative both before and after the pivot.

Of the several forms of the d-step conjecture presented here, the present version is closest to Hirsch's original form [5, pp. 160 and 168] and is
most closely related to linear programming methods. In the example below,
$d = 2$ and the pairs (i,j) under the arrows indicate the positions of the pivot entries.

$$\begin{bmatrix} 1 & 0 & 2 & -1 & 2 \\ 0 & 1 & -1 & 2 & 2 \end{bmatrix} \xrightarrow[(1,3)]{} \begin{bmatrix} 1/2 & 0 & 1 & -1/2 & 1 \\ 1/2 & 1 & 0 & 3/2 & 3 \end{bmatrix} \xrightarrow[(2,4)]{} \begin{bmatrix} 2/3 & 1/3 & 1 & 0 & 2 \\ 1/3 & 2/3 & 0 & 1 & 2 \end{bmatrix}$$
$$\quad\ (I \qquad B \qquad c) \qquad\qquad\qquad\qquad\qquad\qquad\qquad (B' \qquad I \quad c')$$

A set $B \subset R^{d-1}$ is a *simplicial basis* (also called a *minimum positive
basis*) for R^{d-1} if it is the vertex-set of a $(d - 1)$-simplex whose interior
includes the origin. Equivalently, B is affinely independent, is of cardinality d, and the origin 0 is a strictly positive combination of the points
of B. Another equivalent form of the d-step conjecture is reminiscent of the
exchange process used to show all linear bases of a vector space are of the
same cardinality. It is as follows:
If B and B^1 are disjoint simplicial bases of R^{d-1} and the union $U = B \cup B^1$
is a Gaar set (every set of $d - 1$ points of U is linearly independent), then
there is a sequence $B = B_0, B_1, \ldots, N_d = B^1$ of simplicial bases such that for
$1 \leq i \leq d$, B_i is obtained from B_{i-1} by replacing a point of B_{i-1} with a point
of $U \sim B_{i-1}$.
In the example below, $d = 3$ and $0 < \varepsilon < 1$. The rows represent points of B_i.

$$\begin{bmatrix} 1 & 0 \\ 0 & 1 \\ -1 & -1 \end{bmatrix} \longrightarrow \begin{bmatrix} 1 & 0 \\ 0 & 1 \\ -\varepsilon & -1 \end{bmatrix} \longrightarrow \begin{bmatrix} 1 & 1-\varepsilon \\ 0 & 1 \\ -\varepsilon & -1 \end{bmatrix} \longrightarrow \begin{bmatrix} 1 & 1-\varepsilon \\ -1 & -\varepsilon \\ -\varepsilon & -1 \end{bmatrix}$$
$$\quad (B_0 = B) \qquad\qquad (B_1) \qquad\qquad\quad (B_2) \qquad\qquad\quad (B_3 = B')$$

As evidence in favor of the d-step conjecture, one might consider the facts that it is obvious when $d \in \{2,3\}$, is easily proved when $d = 4$ [13], and has been proved for $d = 5$ [15]. In fact, the result for $d = 5$ has been extended by Adler and Dantzig [1] to a much more general class of combinatorial structures. Also, the conjecture has been proved, for arbitrary d, for polytopes arising from certain sorts of linear programs (see [15] for references, and see especially Provan and Billera [22]).

As evidence against the d-step conjecture, we note that when stated without the boundedness condition (b), it is correct when $d \in \{2,3\}$ but not when $d = 4$ [17]. Other strengthened forms of the conjecture have been disproved by Walkup [25], Mani and Walkup [19], and Todd [24].

It seems likely that the conjecture is false for $d = 12$ and perhaps even for $d = 6$. If that is so, what can be said about the asymptotic behavior of $\Delta(d,2d)$ as $d \to \infty$? Does $\Delta(d,2d)$ increase linearly with d? (The known counterexamples [19, 24, 25] to strengthened forms of the conjecture seem to be only "linearly bad.") Quadratically? Polynomially? Exponentially? Any of these conclusions would be of great interest. If it could be shown that $\Delta(d,2d)$ is bounded above by a polynomial in d, the resulting insight might lead to a new pivot rule for the simplex method of linear programming which would combine the practical advantages of Dantzig's pivot rule with the theoretical advantages of the Shor-Khachian algorithm. (Dantzig's algorithm is excellent in the practical sense [5, 7], but its worst-case behavior is exponentially bad [16]. The Shor-Khachian algorithm is "good" in the sense of being polynomially bounded [12], but is not a useful practical tool in its present form [7].) If it could be shown that $\Delta(d,2d)$ increases exponentially with d, that would indicate a strong limitation on the worst-case efficiency of any edge-following algorithm for linear programming.

Though sharper results are known for a few small values of d and n − d (see [15] for references), the best general bounds on $\Delta(d,n)$ are the following, due respectively to Adler [1] and Larman [18]:

$$\left[(n - d) - \frac{(n - d)}{\lceil 5d/4 \rceil} \right] + 1 \leq \Delta(d,n) \leq 2^{d-3}n$$

In particular, $d \leq \Delta(d,2d) \leq 2^{d-3}d$.

REFERENCES

1. I. Adler. Lower bounds for maximum diameters of polytopes, *Math. Programming Study* 1(1974), 11-19.

2. I. Adler and G. B. Dantzig. Maximum diameter of abstract polytopes, *Math. Programming Study* 1(1974), 20-40.

3. D. W. Barnette. Path problems and extremal problems for convex polytopes. Relations between Combinatorics and Other Parts of Mathematics (D. K. Ray Chaudhuri, ed.), *Amer. Math. Soc. Proc. Symp. Pure Math.* 34(1979), 25-34.

4. L. Billera and C. Lee. Sufficiency of McMullen's conditions for f-vectors of simplicial polytopes, *Bull. Amer. Math. Soc.* 2(1980), 181-185.

5. G. B. Dantzig. *Linear Programming and Extensions.* Princeton University Press, Princeton, N. J., 1963.

6. G. B. Dantzig. Eight unsolved problems from mathematical programming, *Bull. Amer. Math. Soc.* 70(1964), 499-500.

7. G. B. Dantzig. Comments on Khachian's algorithm for linear programming, Tech. Report SOL 79-22, Dept. of Operations Research, Stanford University, 1979.

8. B. Grünbaum. *Convex Polytopes.* Pure and Appl. Math., Vol. 16, Interscience, New York, 1967.

9. B. Grünbaum. Polytopes, graphs and complexes, *Bull. Amer. Math. Soc.* 76(1970), 1131-1201.

10. B. Grünbaum. Polytopal graphs, Studies in Graph Theory, Part II (D. R. Fulkerson, ed.), *Math. Assoc. Amer. Studies in Math.* 12(1975), 201-224.

11. B. Grünbaum and G. C. Shephard. Convex polytopes, *Bull. London Math. Soc.* 1(1969), 257-300.

12. L. G. Khachian. A polynomial algorithm in linear programming, *Soviet Math. Doklady* 20(1979), 191-194. (Translated from *Dokl. Akad. Nauk SSSR* 244(1979), 1093-1096.)

13. V. Klee. Diameters of polyhedral graphs, *Canad. J. Math.* 16(1964), 602-614.

14. V. Klee. Convex polytopes and linear programming, *Proc. IBM Sci. Comput. Sympos. Combinatorial Problems*, Yorktown Heights, N. Y., 1964, IBM Data Process Division, White Plains, N. Y., 1966, 123-158.

15. V. Klee. Convex polyhedra and mathematical programming, *Proc. International Congress of Mathematicians*, Vancouver, Canada, 1974, Canadian Math. Congress, 1975, 485-490.

16. V. Klee and G. J. Minty. How good is the simplex algorithm? *Inequalities III* (O. Shisha, ed.), Academic Press, N. Y., 1972, 159-175.

17. V. Klee and D. W. Walkup. The d-step conjecture for polyhedra of dimension d < 6, *Acta Math.* 117(1967), 53-78.

18. D. G. Larman. Paths on polytopes, *Proc. London Math. Soc.* (3)20(1970), 161-178.

19. P. Mani and D. W. Walkup. A 3-sphere counterexample to the W_v-path conjecture, *Math. of Operations Res.* 5(1980), 595-598.

20. P. McMullen. The numbers of faces of simplicial polytopes, *Israel J. Math.* 9(1971), 559-570.

21. P. McMullen and G. C. Shephard. *Convex Polytopes and the Upper Bound Conjecture.* London Math. Soc. Lecture Note Series, 3, Cambridge University Press, London, 1971.

22. J. S. Provan and L. J. Billera. Decompositions of simplicial complexes related to diameters of convex polyhedra, *Math. of Operations Res.* 5 (1980), 576-594.

23. R. Stanley. The number of faces of a simplicial convex polytope, *Advances in Math.* 35(1980), 236-238.

24. M. J. Todd. The monotonic bounded Hirsch conjecture is false for dimension at least 4, *Math. of Operations Res.* 5(1980), 599-601.

25. D. W. Walkup. The Hirsch conjecture fails for triangulated 27-spheres, *Math. of Operations Res.* 3(1978), 224-230.

POLYHEDRAL MAPS ON 2-MANIFOLDS

David Barnette

Department of Mathematics
University of California at Davis
Davis, California

1. INTRODUCTION

Graphs embedded in 2-dimensional manifolds have been studied for about 100 years. Properties involving embeddings and colorings of graphs have been extensively studied. There are, however, a number of problems similar to ones considered for convex polyhedra which have not been investigated. For example, a theorem of Steinitz [8] tells us that any map on the sphere which is 3-connected is isomorphic to the graph of some convex 3-dimensional polytope (hereafter to be called *3-polytopes*). If a map is drawn on some other 2-dimensional manifold, very little is known about when there exists a polyhedron-like structure isomorphic to it. Questions like this will be considered here and properties of such polyhedron-like structures will be examined.

2. POLYHEDRAL MANIFOLDS

Most of the structures to be considered are examples of 2-cell complexes. A *2-cell complex* is a collection of convex polygons such that the intersection of any two is a vertex of both, an edge of both, or is empty. The polygons will be called the *faces* (or sometimes the *facets*) of the complex.

7

A *toroidal polytope* is a 2-cell complex whose union is a torus such that no two faces meeting on an edge are coplanar. A polyhedral 2-manifold is a 2-cell complex whose union is a 2-manifold (embedded in some Euclidean space) such that no two faces meeting on an edge are coplanar.

Two examples of toroidal polytopes are the *triangular picture frame* (Fig. 1) and the famous *Császár Polyhedron* which is a toroidal polytope with seven

FIGURE 1.

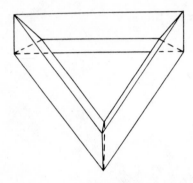

vertices, each two joined by an edge. (For instructions on how to build one, see *Excursions into Mathematics* by Beck, Bleicher, and Crow, Chapter 1.)

We shall use the standard method of representing toroidal maps as maps drawn on a rectangle with the top and bottom to be identified and also the two sides identified. The graphs of these two examples are shown in Figs. 2 and 3, respectively.

FIGURE 2.

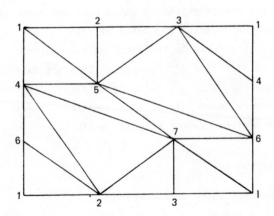

FIGURE 3.

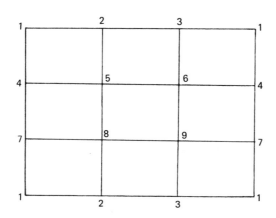

The 3-connectivity of maps on the sphere is equivalent to the property that no two faces have a multiply connected union. For maps on the sphere this is sufficient to guarantee a corresponding polytope. For toroidal maps we shall see that this is not sufficient. It is clearly a necessary condition because two convex polygons cannot have a multiply connected union. We shall say that a map on a 2-manifold is *polyhedral* provided each face of the map is a 2-cell and no two faces have a multiply connected union.

It will be useful to be able to talk about polyhedra that have self intersections, so we shall distinguish between an embedding and an immersion of our polyhedron in some Euclidean space. An *embedding* of a polyhedral map M is a homeomorphic image of M in E^n such that the image of each face of the map is a convex polygon and no two polygons meeting on an edge are coplanar. Embeddings will also be called *realizations* of the map. An *immersion* of a polyhedral map is a continuous image of M in E^n such that each face of M is taken one-to-one onto a convex polygon and no two polygons meeting on an edge are coplanar. In other words, an embedding is a polyhedral manifold, while an immersion is essentially a polyhedral manifold with self intersections.

To see that polyhedrality is not a sufficient condition to guarantee that a toroidal map be realizable, a very large family of such maps will be proved to have no immersions in any Euclidean space.

THEOREM. There are no polyhedral immersions of 3-valent polyhedral maps on orientable manifolds of genus g > 0.

Proof. Let s be the sum of the two-dimensional angles of the faces at their vertices. We use a normalized angle measure in which a 360° angle has a measure of 1. The sum of the angles of an n-gon is thus $(n - 2)/2$. If we let p_i be the number of i-sided faces of any such immersion, then $s = \Sigma(i - 2)p_i/2$ which equals $\frac{1}{2}\Sigma ip_i - \Sigma p_i$ which is just $E - F$, where E and F are the numbers of edges and faces, respectively. Another way to get s is to add the angles around each vertex, doing so for each vertex. Since each vertex is 3-valent, the sum of the angles at the vertex is less than 1, and $s < V$, where V is the number of vertices. We now have that $V > E - F$. Combining this with Euler's equation, $V - E + F = 2 - 2g$, we get $g < 1$, thus, $g = 0$.

This argument clearly holds for all nonorientable 2-manifolds except the projective plane. The theorem is also true for the projective plane but a different proof seems to be necessary.

We now turn to an interesting example to be called the *twisted triangular picture frame* (Fig. 4). It resembles the triangular picture frame,

FIGURE 4.

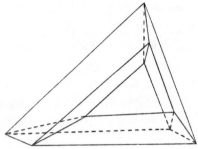

but upon examining it one will see that it does indeed have a twist in it.

When the author's thesis student, Jean Simutis, was working on problems of realizing toroidal maps as polyhedra, he drew her a picture of a twisted triangular picture frame and asked her to give a geometric construction of one. The author felt at the time that from it and the triangular picture frame one could construct a great number of the possible toroidal polytopes. Simutis was very slow to provide the required construction. Even after the author made suggestions on how to begin she couldn't seem to carry it through. He became rather impatient with her. It is almost clear from the picture how to construct it! We finally discovered why she was having so much trouble when she proved that it didn't exist (as a toroidal polytope). Another proof (in fact two proofs) were done independently by a student at Stanford University named Scott Kim. Kim also proved that the 4-sided twisted triangular picture frame and the 4-sided quadrilateral picture frame are realizable (Figs. 5 and 6).

FIGURE 5.   FIGURE 6.

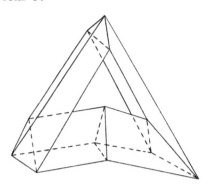

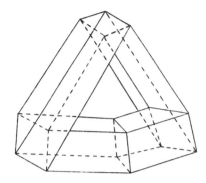

There are very few general realizability theorems for toroidal maps.
One by Altshuler [1] deals with triangulations.

THEOREM. If T is a triangulation of the torus and if T contains a simple
circuit C such that each triangle has an edge on C, then T is realizable as
a toroidal polytope.

There is also a theorem of Simutis [7] which gives us some idea of which
maps are the hardest to show are realizable. We shall let Q be the set of all
maps on the torus such that all faces are 4-sided and all vertices are 4-valent.

THEOREM (Simutis). If M and M' are not in Q, if M' is obtained from M by facet splitting* and if M is realizable, then so it M'.

In her work, Simutis found a striking similarity between a well-known
theorem for 3-polytopes and a property of toroidal polytopes. Using Euler's
equation and a few simple observations and constructions one can show that
the minimum number of edges for a 3-polytope is 6, there is no 3-polytope
with exactly 7 edges, but for all n > 7 there exists a 3-polytope with exactly n edges. Simutis showed that the minimum number of edges of a toroidal
polytope is 18 (the triangular picture frame), there is no toroidal polytope
with exactly 19 edges but for all n > 19 there is a toroidal polytope with exactly n edges.

---

*Facet splitting is a process of adding an edge across a face to obtain a
new map with one more face than before.

We shall now list some open questions dealing with realizability of polyhedral maps.

1. Are there any other neighborly polyhedra besides the tetrahedron and the
    Császár polyhedron? (The modifier *neighborly* means that each two vertices
    are joined by an edge.) Easy calculations using Euler's equation show that
    the simplest unsolved case is the triangulation of the orientable surface
    of genus 6, having 12 vertices.

2. What is "Steinitz's" theorem for tori? (i.e., what are necessary and suf-
    ficient conditions for a toroidal map to be realizable?)

3. Which twisted picture frames are realizable? For each m and n greater
    than 3 there is an m-sided n-gonal picture frame. Furthermore, there are
    [n/2] different types of twists that the picture frame can be given.

4. (Conway). Is there a toroidal polytope that can only be realized in a
    knotted fashion? Conway posed this question about 15 years ago and stated
    that he believed that we were almost to the point where we could answer
    such a question.

5. Are the triangulations of the torus all realizable in E^3? We list this
    as unsolved although Peter Mani claims to have proved that they are all
    realizable. He made this claim seven or eight years ago and a written
    version is yet to be seen.

6. Are the triangulations of the other orientable manifolds all realizable
    in E^3?

7. Are the triangulations of all 2-manifolds realizable in E^4? Since every
    simplicial 2-complex is a subcomplex of a Schlegel diagram of some 6-
    polytope, it follows that they are all realizable in E^5. For the ori-
    entable manifolds the question is whether they can be realized in E^3 or
    E^4. Since the nonorientable manifolds are not topologically realizable
    in E^3, the only hope is for realizations in E^4.

8. What about manifolds of dimensions greater than 2? This is a rather
    vague question because part of the problem is to determine what ques-
    tions should be asked about them.

With regard to Question 7, it can be shown that the triangulations of
the projective plane are realizable in E^4 [4]. The idea is to decompose
the triangulation into a Schlegel diagram of a 3-polytope and a Möbius
strip, then show that the Möbius strip can be realized in E^4 with its vertices arbitrarily close to the corresponding vertices of a polytope isomorphic to the Schelgel diagram. Once this is done it is easy to move the
vertices of the polytope to effect a gluing of the two pieces, producing
the desired polyhedral manifold in E^4.

3. CONVEX AND NONCONVEX VERTICES

In this section we shall deal only with orientable manifolds. We shall say that a vertex of a polyhedral manifold in E^3 *has a vertex figure* provided there is a plane that separates the vertex from its neighbors (i.e., from the vertices to which it is joined by edges). If a vertex has a vertex figure we define the *vertex figure* to be the intersection of such a separating plane with the faces that meet the vertex. A vertex is *convex* provided it has a convex vertex figure.

It is clear that a convex 3-polytope has only convex vertices. It is almost clear that polyhedral manifolds of higher genus must have nonconvex vertices. This can be seen by observing that if all vertices are convex then the sum of the angles of the faces at a vertex is less than 1. The argument that showed that there are no 3-valent immersions now can be carried through to show that there are no polyhedral manifolds with all vertices convex for genus greater than 0. (Since we are now treating realizations in E^3, the projective plane, which is a case not covered in the argument, does not have to be considered.)

Another way to see that there must be nonconvex vertices is to observe that orientable surfaces of genus greater than 0 must have saddle points with respect to some direction. Saddle points can be shown to be nonconvex vertices. It would seem that for a torus there should be at least two saddle points. (Stick a pencil through the hole. Move it up as far as you can and the motion of the pencil should eventually be obstructed by a saddle point. Similarly a saddle point should be found by moving the pencil downward.) It also seems that as the genus increases, the number of saddle points should increase; after all, there are more holes to stick your pencil through.

All this is true if you are dealing with smooth manifolds, but with polyhedral manifolds strange things happen. Banchoff [2] has constructed a toroidal polytope with only one saddle point in one particular direction (in other directions there are more saddle points). Even more incredibly, he constructed orientable polyhedral manifolds for every positive genus, each with only one saddle point in one particular direction.

This doesn't settle the question of how many nonconvex vertices polyhedral manifolds must have. It shows that searching for saddle points probably isn't the way to go about it.

Another problem the author gave to Simutis was to prove that every toroidal polytope has at least six nonconvex vertices. Just as she did with the twisted triangular picture frame, she had difficulty proving it. The author provided her with a heuristic argument: A torus has a hole. A hole has an entrance and an exit. Around an entrance one should get at least three nonconvex vertices and similarly, three more around the exit.

She seemed to have difficulty making this argument work and finally settled for proving that there must be at least three nonconvex vertices in a toroidal polytope.

The author has since proved that there are at least four nonconvex vertices and has constructed a torus with nine vertices that has only five nonconvex vertices (see [3]).

The author has also constructed polyhedral manifolds of all genuses with exactly seven nonconvex vertices (see [3]). The construction is quite similar to Banchoff's.

The edges of polyhedral manifolds in E^3 are of two types. We shall say that an edge is *concave* provided a small disc lying inside the manifold locally supports the edge (note that the manifold separates E^3 into an inside and an outside region). An edge that is not concave is called *convex*. A nonconvex vertex of an embedding of an orientable polyhedral manifold will have both convex and concave edges meeting it. This is not true, however, for immersions.

Since every toroidal polytope has nonconvex vertices, it is clear that every toroidal polytope has concave edges. In fact every one must have at least two concave edges because there are at least four nonconvex vertices. On the other hand there is an embedding of the triangular picture frame with only three concave edges. It is not known if every toroidal polytope has at least three concave edges.

We conjecture that every toroidal polytope has at least three concave edges and that as the genus of an orientable polyhedral manifold embedded in E^3 increases, the minimum number of concave edges it can have will also increase. For immersions, however, the author conjectures that there is a bound on the minimum number of concave edges independent of the genus.

Here are some more unsolved problems:

9. Must all vertices of any realization of the Császár polyhedron be nonconvex?

10. Are there realizations of the Császár polyhedron that are "different" from the usual one?

11. What do we mean by "different realizations" of two polyhedral manifolds?

12. Are there realizable maps on the torus that have no "inside out" reali-
     zation? An example of what is meant by an inside out realization is the
     four-sided triangular picture frame which is an inside out realization
     of the three-sided quadrilateral picture frame.

13. Does there exist a combinatorial type of polyhedral manifold such that
     every realization must have a convex vertex of valence ≥ 4 ?

14. What goes on in higher dimensions? How would we classify the convex
     nature of faces of various dimensions? How many of the various types
     must occur in the various types of manifolds?

15. Is there some kind of relation (perhaps Euler type) between the number
     of convex vertices, nonconvex vertices, concave edges, etc.?

4. GENERATING MAPS AND TRIANGULATIONS

A theorem of Steinitz [8] implies that the triangulations of the two-
sphere can be generated from the boundary of the tetrahedron by a process
called vertex splitting. In vertex splitting, a vertex and two of its incident edges are chosen. The vertex is replaced by two vertices and the
two edges are replaced by two triangles as in Fig. 7.

FIGURE 7.

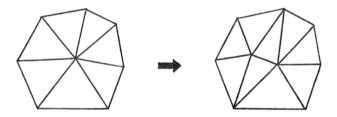

A number of years ago Grünbaum and Duke worked on generating the triangulations of the torus. They found a set of 22 triangulations of the torus
from which one should be able to generate all others by applying sequences of
vertex splittings.

The inverse of vertex splitting is called *edge shrinking*. The 22 triangulations of Grünbaum and Duke were thus minimal with respect to edge
shrinking. The generating problem for triangulations of 2-manifolds is
really the problem of finding the minimal triangulations.

What Grünbaum and Duke wished to do was to use these minimal triangulations to show that all triangulations of the torus were realizable as toroidal polytopes. They planned to show that all minimal triangulations are
realizable, and that realizations of the other triangulations could be

constructed from them by doing a kind of "geometric" vertex splitting. They
were never able to show how the "geometric" vertex splittings could be done
and they finally abandoned the project. The minimal triangulations remained
ignored in Grünbaum's files until a few months ago when this author asked his
thesis student, Kurt Rusnak, to see if he could give a proof that this was the
complete set of minimal triangulations (Grünbaum and Duke never wrote a proof).
He soon found two more minimal triangulations and he seems to have proved that
he has found all of them. Thus, we have:

THEOREM (Grünbaum, Duke, Rusnak). The triangulations of the torus can be
generated from a set of 24 minimal triangulations by vertex splitting.

The generating problem has been solved for one other manifold. The author has recently shown [5] that the triangulations of the projective plane
can be generated from the following triangulations (Fig. 8).

FIGURE 8.

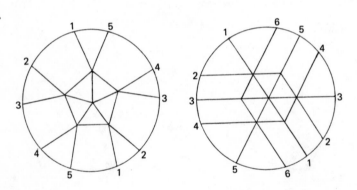

FIGURE 9.

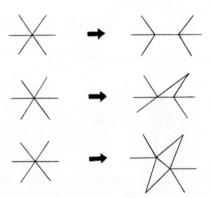

One can also consider generating all polyhedral maps of a manifold. By
using a more general family of vertex splittings (see Fig. 9), one can generate all polyhedral maps on the 2-sphere starting with the boundary of the
tetrahedron [8]. For the projective plane and the torus it appears that the
set of minimal maps will be very large. However, if we admit two processes,
vertex splitting and its dual, facet splitting, then the set of minimal maps
is much smaller. For the projective plane we believe that the following is
the set of minimal maps (Fig. 10).

FIGURE 10.

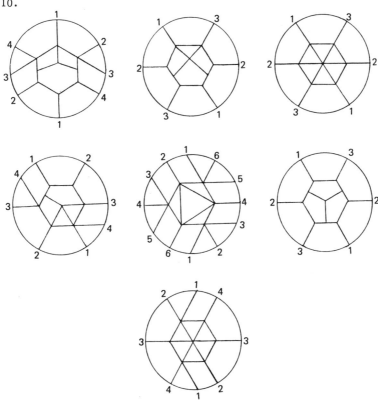

We have the following unsolved problems on generating maps:

16. How does one generate the polyhedral maps on the torus?

17. How does one generate triangulations or polyhedral maps on manifolds
     of higher genus?

18. For every type of 2-manifold is the set of minimal triangulations and
     the set of minimal maps with respect to edge shrinking finite?

19. How does one generate other types of maps such as 2-cell embeddings?

5. A NEW FOUR-COLOR CONJECTURE

Before the four-color theorem was proved, it was well-known that the four-color conjecture was true if and only if it was true for convex 3-polytopes. It was also well-known that maps on the torus are 7-colorable and that some maps require seven colors. It is a little surprising that toroidal polytopes are 6-colorable.

One can prove the 6-coloring property by showing that in the dual graph one can successfully remove vertices of valence at most five until there are at most six vertices left, then coloring the six vertices and returning the vertices one at a time choosing a color for each as it is returned. The author has shown that if there is a polytope for which this is impossible then there exists an immersion of a polyhedral map with all vertices 3-valent [6].

We have not been able to find a toroidal polytope that actually requires six colors. In fact, we cannot find one that requires five colors, and we conjecture that all are 4-colorable. This, however, is not the author's 4-color conjecture. We conjecture that all polyhedral manifolds are 4-colorable.

It hardly seems that this conjecture could be true, yet we can find no counterexample. One difficulty is that we cannot find any polyhedral manifolds for which all faces have a large number of edges.

We conclude with two last questions:

20. Does every polyhedral manifold have a face with five or fewer edges?

21. Does there exist any n such that every polyhedral manifold must have a face with n or fewer edges?

REFERENCES

1. A. Altshuler. Polyhedral realizations in R^3 of triangulations of the torus and 2-manifolds in cyclic 4-polytopes, *Discrete Math.* 1(1971), 211-238.

2. T. F. Banchoff. Critical points and curvature for embedded polyhedral surfaces, *Am. Math. Monthly* 77(1970), 475-485.

3. D. Barnette. Nonconvex vertices of toroidal polytopes, to be published in *Israel J. Math.*

4. D. Barnette. Realizing triangulations of the projective plane in E^4, unpublished.

5. D. Barnette. Generating triangulations of the projective plane, unpublished.

6. D. Barnette. Coloring polyhedral manifolds, unpublished.

7. J. Simutis. Geometric realizations of toroidal maps, Ph.D. Thesis, University of California, Davis, 1977.

8. E. Steinitz and H. Rademacher. *Vorlesungen über die theorie der polyeder*, Berlin, Springer-Verlag, 1934.

CHARACTERIZING THE NUMBERS OF FACES
OF A SIMPLICIAL CONVEX POLYTOPE

Carl W. Lee

Center for Applied Mathematics
Cornell University
Ithaca, New York

1. INTRODUCTION

Polyhedra have long been an object of study, but in the last few decades both the development of linear programming and the expansion of the field of combinatorics have awakened deep interest in d-dimensional polyhedra. Of the many combinatorial questions one may ask about polyhedra, we will single out one: how many faces of various dimensions can a polytope, particularly a simplicial polytope, have? We will recount highlights of endeavors to answer this question, while remaining somewhat faithful to historical chronology.

A *convex polyhedron* is the intersection of a finite number of closed half-spaces in R^d. By a *d-polyhedron* we mean a convex polyhedron of dimension $d \geq 1$. A *convex polytope* is a bounded convex polyhedron; equivalently, it is the convex hull of a finite number of points in R^d. For d-polyhedron P let $f_j(P) = f_j$ denote the number of j-dimensional faces of P, $0 \leq j \leq d - 1$. We shall use the convention $f_{-1}(P) = 1$ throughout this paper. The d-vector $f = f(P) = (f_0(P),\ldots,f_{d-1}(P))$ is the *f-vector* of P. Faces of P of dimension 0, 1 and d - 1 we will call *vertices*, *edges* and *facets*, respectively. A polytope P is *simplicial* if every face of P is a simplex; i.e., if every j-dimensional face of P contains exactly j + 1 vertices of P. Grünbaum [18] may be used for

Current affiliation: Department of Mathematical Sciences, IBM T.J. Watson Research Center, Yorktown Heights, New York; and Department of Mathematical Sciences, University of Kentucky, Lexington, Kentucky

referred to for details of notions in the theory of polytopes and major results in this field through 1967.

Let P^d be the class of all d-polytopes and let P_s^d be the class of all
simplicial d-polytopes, and define

$$f(P^d) = \{f(P) : P \in P^d\}$$

and

$$f(P_s^d) = \{f(P) : P \in P_s^d\}$$

We will concentrate primarily upon $f(P_s^d)$. The recent complete characterization of this set is a major step toward the problem of describing the
f-vectors of all polyhedra, and more generally toward the even more chal-
lenging task of the classification of the combinatorial types of polyhedra.

There are several advantages to focusing attention on simplicial polytopes. They are a "natural" class in the sense that "in general" no more
than d + 1 points chosen from R^d lie on a common hyperplane. Secondly, the
class of simplicial d-polytopes is dual to the class of *simple* d-polytopes
--those with the property that every vertex is on precisely d facets--relevant
to the theory of linear programming. Thirdly, many problems about arbitrary
polytopes, such as the Upper Bound Theorem, are reducible to ones concerning
simplicial polytopes. Finally, the face lattice of a simplicial polytope can
be examined within the more general context of simplicial complexes, allowing
the possibility of the application of algebraic techniques [18, §4.5].

2. EULER'S RELATION

Euler's discovery in 1752 of the relation $f_0 - f_1 + f_2 = 2$ for 3-polytopes
[16, 17] has been acclaimed as "the first important event in topology"
(Alexandroff-Hopf [1, p.1]) and as "the first landmark" in the theory of
polytopes (Klee [21]). The following theorem provides the generalization
of this relation to higher dimensions.

THEOREM 1 (Euler's Relation). If f is in $f(P^d)$ then

$$\sum_{j=0}^{d-1} (-1)^j f_j = 1 + (-1)^{d-1}$$

Schläfli [38] formulated Euler's Relation for d > 3 in 1852 and Poincaré [36] in 1899 provided the first real proof of this generalization. Grünbaum [18, §8.2] introduced in 1967 a completely elementary, nonalgebraic demonstration of Theorem 1. In [18, §8.6] Grünbaum sketches the history of investigations into Euler's Relation and explains that research into its applicability and extensions helped direct attention to the idea of convexity. Imre Lakatos [26] chooses the interesting evolution of Euler's Relation to illustrate his philosophy of the process of mathematical discovery.

It is instructive to observe that quite a few of the early attempts to prove the theorem assumed that the boundary complex of a d-polytope P is *shellable*, that the facets of P can be ordered $F_1, F_2, \ldots, F_m$ so that $F_k \cap (\cup_{1 \leq i \leq k-1} F_i)$ is homeomorphic to a (d − 2)-ball, $2 \leq k \leq m - 1$. This assertion was not established, however, until 1971 by Bruggesser and Mani [13] and has proven to be a powerful tool in the study of polytopes.

3. THE DEHN-SOMMERVILLE EQUATIONS

In terms of linear relations satisfied by all f-vectors of d-polytopes, Euler's Relation is the best possible. Restriction to the class of simplicial polytopes, on the other hand, allows a significant strengthening of Theorem 1, offering the first evidence that $f(P_s^d)$ is a more tractable set than $f(P^d)$.

THEOREM 2 (The Dehn-Sommerville Equations). If f is in $f(P_s^d)$ then

$$(E_k^d) \quad \sum_{j=k}^{d-1} (-1)^j \binom{j+1}{k+1} f_j = (-1)^{d-1} f_k, \qquad -1 \leq k \leq d - 2$$

It is easy to see that (E_{-1}^d) is Euler's Relation and it has been shown that exactly [(d + 1)/2] of the equations (E_k^d) are independent, where [·] denotes the greatest integer function.

In 1905, Dehn [15] worked on the relations for d = 4 and d = 5 and conjectured the existence of analogous relations for d > 5. Sommerville [39] derived the complete system for arbitrary d in 1927. Klee [20] in 1964 rediscovered the Dehn-Sommerville equations in the more general setting of manifolds and incidence systems. In addition to boundary complexes of simplicial polytopes, Theorem 2 applies also to triangulations of topological and homology (d − 1)-spheres and Klee's Eulerian (d − 1)-spheres. See [18, Ch. 9] for more historical details and generalizations.

The next stage in the development of the Dehn-Sommerville equations was their recasting in an especially useful manner by McMullen and Walkup [33, §5.1, 34] in 1971, simultaneously introducing the important notion of the h-vector of a simplicial polytope (g-vector in the original terminology).

Given the d-vector $f = (f_0, f_1, \ldots, f_{d-1})$, define the polynomial

$$f(t) = \sum_{j=-1}^{d-1} f_j t^{j+1}$$

where again by convention we put $f_{-1} = 1$. Now let

$$h(t) = (1 - t)^d f(\frac{t}{1 - t})$$

Then the $(d + 1)$-vector $h = h(f) = (h_0, h_1, \ldots, h_d)$ is defined by the polynomial relation

$$h(t) = \sum_{i=0}^{d} h_i t^i$$

McMullen-Walkup write g_{i-1} instead of h_i. When $f = f(P)$ for some simplicial d-polytope P we call $h(f)$ the *h-vector* of P and denote it by $h(P)$. In this case we also use the natural notation $h_i(P)$ for h_i.

From the definition of $h(f)$ one can explicitly write the h_i as linear combinations of the f_j:

$$h_i = \sum_{j=0}^{i} \binom{d - j}{d - i} (-1)^{i-j} f_{j-1}, \qquad 0 \le i \le d$$

Note that $h_0 = 1$ and $h_1 = f_0 - d$. Further, f can be recovered from $h(f)$ using the relationship

$$f(t) = (1 + t)^d h(\frac{t}{1 + t})$$

or explicitly by

$$f_j = \sum_{i=0}^{j+1} \binom{d - i}{d - j - 1} h_i, \qquad 0 \le j \le d - 1$$

Hence, we have a bijection between the f-vectors $f(P_s^d)$ and the h-vectors $h(P_s^d)$ of simplicial d-polytopes.

In terms of $h(P)$ the Dehn-Sommerville equations become

THEOREM 3. If h is in $h(P_s^d)$ then $h_i = h_{d-i}$, $0 \leq i \leq d$.

So the symmetry of h(P) concisely represents the Dehn-Sommerville equations
and their degree of independence. In this formulation Euler's Relation is
equivalent to $h_0 = h_d$.

The h-vector seems to be a more useful and natural object for consideration than the f-vector for several reasons:

1. As above, h(P) neatly represents the Dehn-Sommerville equations.

2. Since the f_j are nonnegative linear combinations of the h_i, upper and
    lower bounds on the h_i imply upper and lower bounds on the f_j (see
    Sections 5, 6, 8 below).

3. The h-vector precisely captures the changes in the f-vector during the
    process of "shelling" a simplicial complex (Section 4 below).

4. The h-vector has more algebraic significance than the f-vector, as
    abundantly demonstrated by Stanley (Sections 5, 7 below).

5. The complete characterization of $f(P_s^d)$ is actually accomplished in terms
    of $h(P_s^d)$ (Section 7 below).

4. SIMPLICIAL COMPLEXES

We pause here to clarify statement 3. of §3. Δ is said to be a *(finite) simplicial complex* on the finite set V if Δ is a nonempty collection of subsets
of V with the property that $\{v\} \in \Delta$ for all $v \in V$, and that $G \in \Delta$ whenever
$G \subseteq F$ for some $F \in \Delta$. For $F \in \Delta$ we say dim F = j if card F = j + 1, and define dim Δ to be max{dim F : $F \in \Delta$}. For simplicial complex Δ of dimension
$d - 1 \geq 0$ we define $f = f(\Delta) = (f_0(\Delta),\ldots,f_{d-1}(\Delta))$ and $h = h(\Delta) = (h_0(\Delta),\ldots,$
$h_d(\Delta))$ in the obvious way. In particular, $h_d(\Delta) = (-1)^d(1 - \chi(\Delta))$ where
$\chi(\Delta) = \sum_{0 \leq j \leq d-1} (-1)^j f_j(\Delta)$ is the *Euler characteristic* of Δ. If $F \in \Delta$, define the *link* of F in Δ to be the simplicial complex $lk_\Delta F = \{G \in \Delta :$
$F \cap G = \phi, F \cup G \in \Delta\}$. If, further, $F \neq \phi$ define the *deletion* of F from Δ
to be the simplicial complex $\Delta\backslash F = \{G \in \Delta : F \nsubseteq G\}$. Write $|\Delta|$ for the under-
lying topological space of Δ as defined, for example in [40, pp.110-111].

If $F \subseteq V$ we write $\bar{F}$ for the set of all subsets of F. A (d - 1)-dimen-
sional simplicial complex Δ is said to be *(semi)-shellable* if its maximal
faces are all of dimension d - 1 and can be ordered $F_1, F_2, \ldots, F_m$ so that for
$2 \leq k \leq m$

$$\bar{F}_k \cap (\bigcup_{i=1}^{k-1} \bar{F}_i) = \bigcup_{j=1}^{s_k} \bar{G}_j^k$$

where the G_j^k are $s_k \geq 1$ distinct elements of Δ of dimension $d - 2$. For a geometric realization of Δ, this says that the left-hand side above is a topological $(d - 2)$-ball or $(d - 2)$-sphere.

Put $\Delta_k = \cup_{1 \leq i \leq k} \bar{F}_i$, $1 \leq k \leq m$. The relationship between $h(\Delta_{k-1})$ and
$h(\Delta_k)$ is given in [33, §5.2]:

PROPOSITION 4. $h(\Delta_1)$ is the $(d + 1)$-vector $(1,0,0,\ldots,0)$ and for $2 \leq k \leq m$,

$$h_i(\Delta_k) = \begin{cases} h_i(\Delta_{k-1}), & \text{if } i \neq s_k \\ h_i(\Delta_{k-1}) + 1, & \text{if } i = s_k \end{cases}$$

Thus the h-vector increases by one in exactly one component with the addition
of each new facet.

5. THE UPPER BOUND THEOREM

The next major f-vector result was the determination of the maximum number of
j-dimensional faces that a d-polytope with $n > d$ vertices can have. It can be
shown using the procedure of "pulling vertices" of a polytope that from any
nonsimplicial d-polytope P with n vertices one can obtain a simplicial d-
polytope Q with the same number of vertices such that for all $1 \leq j \leq d - 1$,
$f_j(Q) \geq f_j(P)$ [18, §5.2; 33, §2.5]. Hence, it suffices to solve the maximum problem for simplicial polytopes.

One way to construct a simplicial d-polytope with n vertices $v_1,\ldots,v_n$
and large numbers of faces is as follows: Let $c(t) = (t,t^2,\ldots,t^d)$ be the
moment curve in R^d, $d \geq 2$, and select $n > d$ distinct points $v_i = c(t_i)$ on
this curve, where $t_1 < t_2 < \cdots < t^d$. Let C(n,d) be the polytope which is
the convex hull of $\{v_1,\ldots,v_n\}$. Define C(2,1) to be any 1-simplex. The face
lattice of C(n,d) is actually independent of the particular values of t_i, so
C(n,d) is referred to as *the cyclic d-polytope with n vertices* [18, §4.7;
33, §2.3(vi)]. Now it happens that if $1 \leq k \leq [d/2]$ then the convex hull of
any k-subset of V is a face of C(n,d). Thus, for $0 \leq j \leq [d/2] - 1$, this
polytope clearly has the maximum number of j-dimensional faces of any d-
polytope with n vertices. It is therefore natural to inquire whether C(n,d)
has this property for higher dimensional faces as well. In 1957, Motzkin
[35] (implicitly) conjectured that this is indeed the case.

Attempts to establish the truth of Motzkin's conjecture [18, §10.1; 33] often used a theorem of Kruskal [25], published in 1963, which places bounds on the numbers of faces of arbitrary simplicial complexes. For positive integers r and i, r can be written uniquely in the form

$$r = \binom{n_i}{i} + \binom{n_{i-1}}{i-1} + \cdots + \binom{n_k}{k}$$

where $n_i > n_{i-1} > \cdots > n_k \geq k \geq 1$. This is called the *i-canonical representation of r*. If, in addition, we are given a positive integer j we define

$$r^{\{j|i\}} = \binom{n_i}{j} + \binom{n_{i-1}}{j-1} + \cdots + \binom{n_k}{k+j-i}$$

(We will always use the convention $\binom{a}{0} = 1$ if $a < 0$ and otherwise $\binom{a}{b} = 0$ if $a < b$ or $b < 0$.) For nonnegative r,i,j put

$$\kappa(r,i,j) = \begin{cases} \max\{f_j(\Delta) : f_i(\Delta) = r\}, & \text{if } j > i \\ \min\{f_j(\Delta) : f_i(\Delta) = r\}, & \text{if } j < i \end{cases}$$

as Δ ranges over all simplicial complexes.

THEOREM 5 (Kruskal). For all nonegative i,j and positive r, $\kappa(r,i,j) = r^{\{j+1|i+1\}}$. In fact, a d-vector $(f_0,\ldots,f_{d-1})$ of positive integers is the f-vector of some (d − 1)-dimensional simplicial complex if and only if $f_{j+1} \leq f_j^{\{j+2|j+1\}}$, $0 \leq j \leq d - 2$.

Hence, we have a complete characterization of f-vectors of simplicial complexes.

Motzkin's conjecture became a theorem with McMullen's proof in 1970 [30; 33, Ch. 5]. McMullen did not use Kruskal's theorem; the main ingredients of his proof are (1) the existence of a particular shelling order of boundary complexes of simplicial polytopes and (2) use of the h-vector instead of the f-vector.

A little bit of calculation shows that $h_i(C(n,d)) = h_{d-i}(C(n,d)) = \binom{n - d + i - 1}{i}$, $0 \leq i \leq [d/2]$. McMullen established that for a simplicial d-polytope P with n vertices, $0 \leq h_i(P) = h_{d-i}(P) \leq \binom{n - d + i - 1}{i}$,

$0 \le i \le [d/2]$. The Upper Bound Theorem now follows immediately from the remark that the f_j are nonnegative linear combinations of the h_i.

UPPER BOUND THEOREM 6 (McMullen). If f is in $f(P^d)$ and $f_0 = n$ then
$$f_j \le f_j(C(n,d)), \qquad 0 \le j \le d - 1.$$

For future reference we give here the numbers $f_j(C(n,d))$ explicitly [33, §2.3(vi); 18, §4.7].

$$f_j(C(n,d)) = \begin{cases} \sum_{i=1}^{[d/2]} \frac{n}{n-i} \binom{n-i}{i}\binom{i}{j-i+1}, & 0 \le j \le d-1, \\ & \text{if d is even} \\ \sum_{i=0}^{[d/2]} \frac{j+2}{n-i} \binom{n-i}{i+1}\binom{i+1}{j-i+1}, & 0 \le j \le d-1, \\ & \text{if d is odd} \end{cases}$$

In particular,

$$f_j(C(n,d)) = \binom{n}{j+1}, \qquad 0 \le j \le [d/2] - 1,$$

$$f_{d-1}(C(n,d)) = \binom{n - [(d+1)/2]}{n-d} + \binom{n - [(d+2)/2]}{n-d}$$

Even though no d-polytope has just d vertices, it is also helpful formally to define the numbers

$$f_j(C(d,d)) = \begin{cases} \binom{d}{j+1}, & \text{if } 0 \le j \le d-2 \\ 2, & \text{if } j = d-1 \end{cases}$$

There is an inherent difficulty in adapting McMullen's proof to triangulations of the (d - 1)-sphere; i.e., to simplicial complexes Δ for which |Δ| is a topological (d - 1)-sphere, since not all such triangulations are shellable [14]. However, Stanley [42] in 1975 extended Theorem 6 to topological and homology (d - 1)-spheres as well. The method of proof seems to outweigh the value of the particular result, opening up many new possibilities for the interaction between combinatorics and commutative algebra.

A nonempty set M of monomials $Y_1^{a_1} \cdots Y_s^{a_s}$ in the variables $Y_1, Y_2, \ldots, Y_s$ is an *order ideal of monomials* if it has the property that $m_2 \in M$ whenever $m_2 | m_1$ for some $m_1 \in M$. A finite or infinite sequence of integers $(h_0, h_1, h_2, \ldots)$

is called an *0-sequence* if there exists an order ideal of monomials M such that $h_i = \text{card}\{m \in M : \deg m = i\}$, where deg m is the degree of the monomial m.

Given positive integers h and i define

$$h^{<i>} = \binom{n_i + 1}{i + 1} + \cdots + \binom{n_k + 1}{k + 1}$$

where

$$h = \binom{n_i}{i} + \cdots + \binom{n_k}{k}$$

is the i-canonical representation of h. Also put $0^{<i>} = 0$.

THEOREM 7 (Stanley). Let $(h_0, h_1, \ldots, h_d)$ be a $(d + 1)$-vector of integers. Then the following four conditions are equivalent:

(i) $(h_0, \ldots, h_d)$ is the h-vector of some simplicial $(d - 1)$-complex Δ such that for all $F \in \Delta$, $\tilde{H}_i(\text{lk}_\Delta F; Q) = 0$ if $i \neq \dim \text{lk}_\Delta F$.

(ii) $(h_0, \ldots, h_d)$ is the h-vector of some shellable $(d - 1)$-complex.

(iii) $(h_0, \ldots, h_d)$ is an 0-sequence.

(iv) $h_0 = 1$, $h_1 \geq 0$ and $0 \leq h_{i+1} \leq h_i^{<i>}$ for all $1 \leq i \leq d - 1$.

In (i) above, $\tilde{H}_i(\text{lk}_\Delta F; Q)$ denotes the i-th reduced simplicial homology module of the simplicial complex $\text{lk}_\Delta F$ computed with coefficients in the field Q of rational numbers [37]. Note the similarity between condition (iv) and Kruskal's condition in Theorem 5.

COROLLARY 8. Condition (iv) holds for h-vectors of d-polytopes and of triangulated topological and homology $(d - 1)$-spheres, and as a consequence the Upper Bound Theorem also holds for these objects.

The basic idea is that with a simplicial complex Δ one can associate a particular graded Q-algebra A_Δ. Reisner [37] demonstrated that A_Δ is Cohen-Macaulay if and only if Δ has the property described in (i). Such complexes are therefore called *Cohen-Macaulay complexes*. Stanley explained how in this case one can obtain an order ideal of monomials for which $(h_0(\Delta), \ldots, h_d(\Delta))$ is the corresponding 0-sequence and offered (iv) as an explicit numerical characterization of 0-sequences. For details of this work and some of its extensions see [19, 37, 41-45].

6.   THE LOWER BOUND THEOREM

The determination of the minimum number of j-dimensional faces that a d-polytope P with n vertices can have seems to be considerably harder than the corresponding maximum question, partly because the minima are definitely not a-
chieved by simplicial polytopes.   Some attempts have been made to specify the
minimum value of $f_{d-1}(P)$--for example, see Grünbaum [18, §10.2] and McMullen
[31].

There was, however, a long-standing conjecture as to the minimum numbers
of faces of a simplicial d-polytope with n vertices [18, §10.2].   Using the
terminology of [21] let $P(d,d + 1)$ be a d-simplex, and for $n > d + 1$ let $P(d,n)$
be obtained from $P(d,n - 1)$ by adding a pyramidal cap over one of the facets of
$P(d,n - 1)$.   Then $P(d,n)$ is a simplicial d-polytope with n vertices.   The conjecture was that no simplicial d-polytope with n vertices has fewer faces of
any dimension than $P(d,n)$.   The conjecture for $d = 4$ was first stated by
Brückner [12] in 1909 as a theorem, but his proof was later shown to be in-
valid.   Barnette established the conjecture for $j = d - 1$ in 1971 [2]   and
for $1 \leq j \leq d - 2$ in 1973 [5].

LOWER BOUND THEOREM 9 (Barnette).   If f is in $f(P_s^d)$ and $f_0 = n$, then

$$
f_j \geq \begin{cases} \binom{d}{j + 1} + (n - d)\binom{d}{j}, & \text{if } 1 \leq j \leq d - 2 \\[2ex] (n - d)(d - 1) + 2, & \text{if } j = d - 1 \end{cases}
$$

(For some extensions of Theorem 9 to manifolds and pseudo-manifolds, see [4,
24, 47].)

It is not hard to see that the truth of the Lower Bound Theorem would
follow from a demonstration that $h_i(P) \geq h_1(P) = n - d$, $2 \leq i \leq d - 1$,
for all $P \in P_s^d$ with n vertices.   In 1971, McMullen and Walkup [34] conjectured an even stronger condition on the h-vectors of simplicial polytopes;
namely, that the h-vector is unimodal:

GENERALIZED LOWER BOUND THEOREM 10.   If h is in $h(P_s^d)$ then $h_{i+1} \geq h_i$, $0 \leq$
$i \leq [d/2] - 1$.

They proved that the above condition is the strongest possible in the
sense that any linear inequality satisfied by all the vectors in $f(P_s^d)$ is a
consequence of the inequalities of Theorem 10.   They also showed that the

condition holds for simplicial d-polytopes which can be subdivided into simplicial complexes without introducing any new simplices of dimension less than
$[d/2]$. The conjecture was confirmed for $d \leq 5$ and for $d < n \leq d + 3$ by McMullen [32], and fully established as an immediate consequence of the main theorem of the next section.

7. THE SET OF h-VECTORS OF SIMPLICIAL POLYTOPES

We now arrive at the central theorem of this report: the complete characterization of the h-vectors and, hence, of the f-vectors, of simplicial polytopes.

Of course, there is no difficulty in describing $f(P^2) = f(P_s^2)$. Steinitz
[46] characterized $f(P^3)$ in 1906, and the description of $f(P_s^3)$ follows readily
[18, §10.3]. Some facts about $f(P^4)$ and $f(P_s^4)$ are examined in [3; 6; 7; 18,
§10.4]. Showing remarkable insight, McMullen [32] in 1971 proposed a set of
conditions to characterize $h(P_s^d)$, later established as a theorem, stated as
follows.

THEOREM 11. The vector $h = (h_0, \ldots, h_d)$ of integers is in $h(P_s^d)$ if and only if
the following three conditions hold:

(i) $h_i = h_{d-i}$, $0 \leq i \leq d$ (Dehn-Sommerville equations),

(ii) $h_{i+1} \geq h_i$, $0 \leq i \leq [d/2] - 1$ (Generalized Lower Bound conditions)
       and

(iii) $h_0 = 1$ and $h_{i+1} - h_i \leq (h_i - h_{i-1})^{<i>}$, $1 \leq i \leq [d/2] - 1$.

Together, (ii) and (iii) are equivalent to the statement that $(h_0, h_1 - h_0,$
$\ldots, h_{[d/2]} - h_{[d/2]-1})$ is an O-sequence. McMullen proved his conjecture in
the case $d \leq 5$ and in the case $d < n \leq d + 3$, the latter involving the use of
Gale diagrams [18, §§5.4, 6.3; 33, Ch. 3]. Work of Walkup [47] in 1970 and
Mani [28] in 1972 imply that Theorem 11 also holds for triangulations of $(d - 1)$-spheres in these specific cases.

By algebraic means, Stanley [43, 44] showed in 1977 that (i), (ii) and
(iii) hold for simplicial $(d - 1)$-complexes Δ if A_Δ is Gorenstein, $h_d = 1$ and
$n \leq d + 3$ and, hence, provided another proof of the necessity of (i), (ii)
and (iii) for triangulated spheres with few vertices.

In 1979, Billera and the author [8, 9, 27] derived the sufficiency of
McMullen's conditions during an attempt to strengthen some results of Klee
[23] (see §8). Shortly after, Stanley [45] proved the necessity, thereby
establishing the theorem. Notable in Stanley's proof is the employment of

algebraic geometry and especially the Hard Lefschetz Theorem. Unlike his proof
of the Upper Bound Theorem, Stanley's proof here applies exclusively to boundary complexes of simplicial convex polytopes and not, therefore, to the properly larger class of triangulated spheres. Whether Theorem 11 can be extended
to spheres is still an open question.

The cyclic polytope is the key to the proof of sufficiency. The cyclic
$(d + 1)$-polytope contains, in some sense, a rich collection of simplicial d-
polytopes. For $h \in h(P_s^d)$ with $h_1 = n - d$, a d-polytope P with $h(P) = h$ can be
obtained in the following way: Construct $C(n, d + 1)$ with a "suitable" choice
of $t_1 < t_2 < \cdots < t_n$. Use h to select a certain proper shellable collection
B of facets of $C(n, d + 1)$ and find a point $z \in R^{d+1}$ from which the facets of
the cyclic that are "visible" are precisely those in B [18, §5.2]. Let H
be a hyperplane strictly separating z from the vertices of $C(n, d + 1)$. Then
the desired polytope P will be the intersection of H with the convex hull of
$C(n, d + 1)$ and the point z. Further, P comes with a natural subdivision into
a simplicial complex with no nonboundary simplices of dimension less than
$[(d + 1)/2]$.

The idea of obtaining new polytopes in this manner from a cyclic polytope
was inspired by Klee [23] (see also Grünbaum [18, §7.2]) and the selection procedure for B was motivated by the construction demonstrating (iv) implies (ii)
of Theorem 7.

Theorems 1, 2, 3, 6, 9, and 10 and the fact that condition (iv) of Theorem 7 holds for d-polytopes are all consequences of Theorem 11. For every simple d-polytope P one can find a simplicial d-polytope P* dual to P in the sense
that there is an inclusion-reversing bijection between the set of all faces of
P and the set of all faces of P* [18, §3.4]. So we have with no further effort a characterization of f-vectors of simple d-polytopes, since $(f_0, f_1, \ldots,$
$f_{d-1})$ is the f-vector of a simple d-polytope if and only if $(f_{d-1}, \ldots, f_1, f_0)$
is the f-vector of a simplicial d-polytope. The complete characterization of
$f(P^d)$ remains unknown.

8. POLYTOPE PAIRS AND UNBOUNDED POLYHEDRA

In this section we turn to some of the implications of the previous section
for simplicial polytopes with additional structure and comment on applications
to the numbers of faces of unbounded simple polyhedra. These ideas were studied by Klee [23] and extended in [10, 27].

A polyhedron is *pointed* if it has at least one vertex. Klee [22] proved in 1966 that every simple, pointed d-polyhedron with n facets has at least n − d + 1 vertices. Grünbaum [18, §10.2] speculated whether this result might be improved upon if one specified both the number of bounded and unbounded facets. Problems of this form were successfully approached from the point of view of polytope pairs by Klee [23] in 1974, while investigating the efficiency of a proposed algorithm of Mattheiss [29] to enumerate the vertices of a simple polytope defined by linear inequalities.

By a *polytope pair* (P,F) of type (d,n,u), $3 \leq d \leq u < n$, we mean a simple d-polytope P with n facets, one of which is the simple (d − 1)-polytope F with u facets. Given polytope pair (P,F) of type (d,n,u), applying a projective transformation [18, §1.1] which sends the facet F into the hyperplane at infinity yields $P \sim F$, a simple, pointed d-polyhedron with n − 1 facets, u of which are unbounded. Moreover, every simple, pointed, unbounded d-polyhedron is projectively equivalent to $P \sim F$ for some polytope pair (P,F).

Let (P,F) be a polytope pair of type (d,n,u). Then P is dual to a simplicial d-polytope P* with n vertices, one of which, v_F, is on precisely u edges. Let Δ be the boundary complex of P*. We then have the following relationships between (P,F) and (Δ, v_F):

$$f_j(P) = f_{d-j-1}(\Delta), \qquad 0 \leq j \leq d - 1,$$

$$f_j(F) = f_{d-j-2}(1k_\Delta v_F), \qquad 0 \leq j \leq d - 2 \text{ and}$$

$$f_j(P \sim F) = f_{d-j-1}(\Delta \backslash v_F), \qquad 0 \leq j \leq d - 1$$

Theorem 11 and related constructions can be used in a variety of ways to determine extremal values for the components of the h-vectors and f-vectors of Δ, $1k_\Delta v_F$ and $\Delta \backslash v_F$. We summarize these results in a manner analogous to Klee's format.

THEOREM 12. As (P,F) ranges over all polytope pairs of type (d,n,u), $3 \leq d \leq u < n$, we have the following minima and maxima:

	function	minimum	
(i)	$h_i(\Delta)$	$\begin{cases} 1 \\ n - d \end{cases}$	$i = 0$ or d $1 \leq i \leq d - 1$

(ii)
$$h_i(1k_\Delta v_F) \begin{cases} 1 & i = 0 \text{ or } d - 1 \\ u - d + 1 & 1 \le i \le d - 2 \end{cases}$$

(iii)
$$h_i(\Delta \backslash v_F) \begin{cases} 1 & i = 0 \\ n - d - 1 & i = 1 \\ n - u - 1 & 2 \le i \le d - 1 \\ 0 & i = d \end{cases}$$

(iv)
$$f_j(\Delta) \begin{cases} \binom{d}{j+1} + (n - d)\binom{d}{j} & 0 \le j \le d - 2 \\ (n - d)(d - 1) + 2 & j = d - 1 \end{cases}$$

(v)
$$f_j(1k_\Delta v_F) \begin{cases} \binom{d-1}{j+1} + (u - d + 1)\binom{d-1}{j} & 0 \le j \le d - 3 \\ (u - d + 1)(d - 2) + 2 & j = d - 2 \end{cases}$$

(vi)
$$f_j(\Delta \backslash v_F) \begin{cases} \binom{d}{j+1} + (n-d-1)\binom{d-1}{j} + (n-u-1)\binom{d-1}{j-1} & 0 \le j \le d - 2 \\ (n - u - 1)(d - 2) + n - d & j = d - 1 \end{cases}$$

function   maximum

(vii)
$$h_i(\Delta) \begin{cases} \binom{n-d+i-2}{i} + \binom{u-d+i-1}{i-1} & 0 \le i \le [d/2] \\ \binom{n-i-2}{d-i} + \binom{u-i-1}{d-i-1} & [d/2]+1 \le i \le d \end{cases}$$

(viii)
$$h_i(1k_\Delta v_F) \begin{cases} \binom{u-d+i}{i} & 0 \le i \le [(d-1)/2] \\ \binom{u-i-1}{d-i-1} & [(d-1)/2]+1 \\ & \quad \le i \le d-1 \end{cases}$$

(ix)
$$h_i(\Delta \backslash v_F) \begin{cases} \binom{n-d+i-2}{i} & 0 \le i \le [d/2] \\ \binom{n-i-2}{d-i} & [d/2]+1 \le i \le d-2 \\ n - u - 1 & i = d - 1 \\ 0 & i = d \end{cases}$$

(x) $f_j(\Delta)$ $f_j(C(n-1,d)) + f_j(C(u+1,d))$ $0 \le j \le d-1$

$- f_j(C(u,d))$

(xi) $f_j(1k_\Delta v_F)$ $f_j(C(u,d-1))$ $0 \le j \le d-2$

(xii)

$f_j(\Delta \backslash v_F)$
$$\begin{cases} f_j(C(n-1,d)) & 0 \le j \le d-3 \\ f_{d-2}(C(n-1,d)) + d - u & j = d - 2 \\ f_{d-1}(C(n-1,d)) + d - u - 1 & j = d - 1 \end{cases}$$

All of the above f-vector results can, of course, be recast in a dual fashion for $f_j(P)$, $f_j(F)$ and $f_j(P \sim F)$. In particular, examining $f_j(P \sim F)$ yields:

COROLLARY 13. Let $3 \le d \le u \le n$. As P ranges over all simple, pointed d-polyhedra with n facets, u of which are unbounded, then

(i)
$$\min f_j(P) = \begin{cases} (n-u)(d-2) + n - d + 1, & \text{if } j = 0 \\ \binom{d}{j} + (n-d)\binom{d-1}{j} + (n-u)\binom{d-1}{j+1}, & \text{if } 1 \le j \le d-1 \end{cases}$$

(ii)
$$\max f_j(P) = \begin{cases} f_{d-1}(C(n,d)) + d - u - 1, & \text{if } j = 0 \\ f_{d-2}(C(n,d)) + d - u, & \text{if } j = 1 \\ f_{d-j-1}(C(n,d)), & \text{if } 2 \le j \le d-1 \end{cases}$$

Part (i) of the corollary confirms a conjecture of Björner [11]. Theorem 12 can be further refined if more of the components of $h(1k_\Delta v_F)$ are known and Corollary 13 can be strengthened if the dimension of the recession cone of P is also specified, so it seems quite likely that these types of methods may bring within reach a complete characterization of the set of f-vectors of un-bounded simple polyhedra.

REFERENCES

1.   P. Alexandroff and H. Hopf. *Topologie*, Springer-Verlag, Berlin, 1935.

2.   D. W. Barnette. The minimum number of vertices of a simple polytope, *Israel J. Math.* 10(1971), 121-125.

3.   D. W. Barnette. Inequalities for f-vectors of 4-polytopes, *Israel J. Math.* 11(1972), 284-291.

4.   D. W. Barnette. Graph theorems for manifolds, *Israel J. Math.* 16(1973), 62-72.

5.   D. W. Barnette. A proof of the lower bound conjecture for convex poly-
     topes, *Pacific J. Math.* 46(1973), 349-354.

6.   D. W. Barnette. The projection of the f-vectors of 4-polytopes onto
     the (E,S)-plane, *Discrete Math.* 10(1974), 201-216.

7.   D. W. Barnette and J. R. Reay. Projections of f-vectors of four-poly-
     topes, *J. Combinatorial Theory* 15(1973), 200-209.

8.   L. J. Billera and C. W. Lee. Sufficiency of McMullen's conditions for
     f-vectors of simplicial polytopes, *Bull. Amer. Math. Soc. (New Series)*
     2(1980), 181-185.

9.   L. J. Billera and C. W. Lee. A proof of the sufficiency of McMullen's
     conditions for f-vectors of simplicial convex polytopes, Tech. Rep. No.
     469, School of O. R. and I. E., Cornell University, Ithaca, New York,
     1980.

10. L. J. Billera and C. W. Lee. The numbers of faces of polytope pairs
     and unbounded polyhedra, Research Report, IBM T. J. Watson Research
     Center, Yorktown Heights, New York, unpublished.

11. A. Björner. The minimum number of faces of a simple polyhedron, *Euro-
     pean J. Combinatorics* 1(1980), 27-31.

12. M. Brückner. Ueber die Ableitung der allgemeinen Polytope und die nach
     Isomorphismus verschiedenen Typen der allgemeinen Achtzelle (Oktatope),
     Verhandel. Koninkl. Akad. Wetenschap. (Eerste Sectie), Vol. 10, No. 1
     (1909).

13. H. Bruggesser and P. Mani. Shellable decompositions of cells and spheres,
     Math. Scand. 29(1971), 197-205.

14. G. Danaraj and V. Klee. Which spheres are shellable?, *Ann. Discrete Math.*
     2(1978), 33-52.

15. M. Dehn. Die Eulersche Formel in Zussammenhang mit dem Inhalt in der
     nicht-Euklidischen Geometrie, *Math. Ann.* 61(1905), 561-586.

16. L. Euler. Elemental doctrinae solidorum, *Novi Comm. Acad. Sci. Imp.
     Petropol.* 4(1752/53), 109-140.

17. L. Euler. Demonstratios nunnullarum insignium proprietatum, quibus
     solida hedris planis inclusa sunt praedita, *Novi Comm. Acad. Sci. Imp.
     Petropol.* 4(1752/53), 140-160.

18. B. Grünbaum. *Convex Polytopes,* Wiley, New York, 1967.

19. M. Hochster. Cohen-Macaulay rings, combinatorics, and simplicial com-
     plexes, *Ring Theory II -- Proceedings of the Second Oklahoma Conference*
     (B. R. McDonald and R. A. Morris, eds.), Marcel Dekker, New York, 1977.

20. V. Klee. A combinatorial analogue of Poincaré's duality theorem, *Canad.
     J. Math.* 16(1964), 517-531.

21. V. Klee. Convex polytopes and linear programming, *Proceedings of the
     IBM Scientific Computing Symposium on Combinatorial Problems, March 16-
     18, 1964,* IBM Data Processing Division, White Plains, New York, 1966,
     123-158.

22. V. Klee. A comparison of primal and dual methods for linear programming,
     Numer. Math. 9(1966), 227-235.

23. V. Klee. Polytope pairs and their relationship to linear programming, *Acta Math.* 133(1974), 1–25.

24. V. Klee. A d-pseudomanifold with f_0 vertices has at least $df_0 - (d-1)(d+2)$ d-simplices, *Houston J. Math.* 1(1975),81–86.

25. J. B. Kruskal. The number of simplices in a complex, *Symposium on Mathematical Optimization Techniques, Berkeley 1960,* Berkeley, 1963, 251–278.

26. I. Lakatos. *Proofs and Refutations,* Cambridge U. Press, Cambridge, 1976.

27. C. W. Lee. Counting the faces of simplicial convex polytopes, Ph. D. Thesis, Cornell U., Ithaca, New York, 1981.

28. P. Mani. Spheres with few vertices, *J. Combinatorial Theory (A)* 13(1972), 346–352.

29. T. H. Mattheiss. An algorithm for determining irrelevant constraints and all vertices in systems of linear inequalities, *Operations Res.* 21 (1973), 247–260.

30. P. McMullen. The maximum numbers of faces of a convex polytope, *Mathematika* 17(1970), 179–184.

31. P. McMullen. The minimum number of facets of a convex polytope, *J. London Math. Soc. (2)*3(1971), 350–354.

32. P. McMullen. The numbers of faces of simplicial polytopes, *Israel J. Math.* 9(1971), 559–570.

33. P. McMullen and G. C. Shephard. Convex polytopes and the upper bound conjecture, *London Math. Soc. Lecture Note Series 3,* Cambridge U. Press, Cambridge, 1971.

34. P. McMullen and D. W. Walkup. A generalized lower-bound conjecture for simplicial polytopes, *Mathematika* 18(1971), 264–273.

35. T. S. Motzkin. Comonotone curves and polyhedra, Abstract 111, *Bull. Amer. Math. Soc.* 63(1957), 35.

36. H. Poincaré. Complément a l'analysis situs, *Rend. Circ. Mat. Palermo* 13(1899), 285–343.

37. G. A. Reisner. Cohen-Macaulay quotients of polynomial rings, *Advances in Math.* 21(1976), 30–49.

38. L. Schläfli. Theorie der vielfachen Kontinuität, *Denkschr. Schweiz. naturf. Ges.* 38(1901), 1–237.

39. D. M. Y. Sommerville. The relations connecting the angle-sums and volume of a polytope in space of n dimensions, *Proc. Roy. Soc. London, Ser. A* 115(1927), 103–119.

40. E. H. Spanier. *Algebraic Topology,* McGraw-Hill, New York, 1966.

41. R. P. Stanley. Cohen-Macaulay rings and constructible polytopes, *Bull. Amer. Math. Soc.* 81(1975), 133–135.

42. R. P. Stanley. The upper bound conjecture and Cohen-Macaulay rings, *Stud. in Appl. Math.* 54(1975), 135–142.

43. R. P. Stanley. Cohen-Macaulay complexes, *Higher Combinatorics* (M. Aigner, ed.), D. Reidel, Dordrecht-Holland, 1977, 51–62.

44. R. P. Stanley. Hilbert functions of graded algebras, *Advances in Math.*
     28(1978), 57–83.

45. R. P. Stanley. The number of faces of a simplicial convex polytope, *Advances in Math.* 35(1980), 236–238.

46. E. Steinitz. Über die Eulersche Polyederrelationen, *Arch. Math. Phys.*
     *(3)*11(1906), 86–88.

47. D. W. Walkup. The lower bound conjecture for 3- and 4-manifolds, *Acta Math.* 125(1970), 75–107.

A DUAL PROOF OF THE UPPER BOUND THEOREM

Arne Brøndsted

Institute of Mathematics
University of Copenhagen
Copenhagen, Denmark

The upper bound theorem, proved by P. McMullen [2] in 1970, may be given the following dual formulation:

THEOREM. There exist numbers $\phi_k(v,d)$ such that

$$f_k(Q) \leq \phi_k(v,d), \qquad k = 0,\ldots,d-2 \tag{*}$$

for any simple d-polytope Q with v facets, and

$$f_k(Q) = \phi_k(v,d), \qquad k = 0,\ldots,d-2 \tag{**}$$

for any simple d-polytope Q with v facets which is the dual of a neighborly polytope.

(As usual, $f_k(Q)$ denotes the number of k-faces of Q.) A proof of the upper bound theorem in the dual setting has been given by Aa. Bondesen and A. Brøndsted [1]. In the following, we shall outline that proof. It is an adapted version of McMullen's original proof, containing major simplifications.

A d-polytope Q is *simple* if each (d - k - 1)-face of Q is contained in, and hence is the intersection of, exactly k + 1 facets, k = 0,...,d - 1; in fact, this holds if it holds for k = d - 1. Equivalently, a d-polytope is simple if it is the dual of a simplicial d-polytope. In the proof we need the following elementary facts about a simple d-polytope Q:

   (a) For each k-face G of Q and each vertex x of G there are exactly k
        edges in G containing x, for k = 0,...,d.

   (b) Let $E_1,...,E_k$ be edges in Q, for k = 0,...,d, containing a common
        vertex x. Then the smallest face G of Q containing $E_1,...,E_k$ has
        dimension k, and $E_1,...,E_k$ are the only edges in G containing x.

   (c) Each facet of Q is again a simple polytope.

   (d) Q is the dual of a neighborly polytope if and only if any [d/2] or
        fewer facets of Q have a nonempty intersection.

In the following, let Q be a simple d-polytope in R^d. We shall turn the 1-skeleton of Q into an *oriented graph* as follows. We let w be a vector in R^d such that for any two vertices x and y we have <w,x> ≠ <w,y>. We then agree to orient an edge with endpoints x and y towards x and away from y when <w,x> > <w,y>. Calling the direction of w the "down" direction, this amounts to orienting the edges "downwards."

For a vertex x of Q we define the *in-degree* of x as the number of edges in Q which have x as an endpoint and are oriented towards x. Similarly, the *out-degree* of x is defined as the number of edges which have x as an endpoint and are oriented away from x. Taking G = Q in (a) we see that

   the sum of the in-degree and the out-degree of any vertex   (1)
   x is d.

By a *k-in-star*, k = 0,...,d, we shall mean a set consisting of a vertex x of Q and k edges in Q having x as an endpoint and being oriented towards x. A *k-out-star* is defined similarly. Denoting by γ_k the number of vertices of Q with in-degree k, we see that

$$\text{the number of k-in-stars in Q is } \sum_{j=0}^{d} \binom{j}{k}\gamma_j. \qquad (2)$$

The face-structure of Q is linked to the graph-structure by the observation that there is a one-to-one correspondence between the k-faces of Q and the k-in-stars. In fact, to each k-face G there corresponds in a natural way a k-in-star, namely the "lowest" vertex of G together with the k edges of G which contain x, cf. (a). Conversely, using (b) it is not difficult to see that each k-in-star is contained in a unique k-face G and that the k-in-star is in fact the k-in-star corresponding to G as described above. Denoting by f_k the number of k-faces of Q, it then follows from (2) that we have

$$f_k = \sum_{j=0}^{d} \binom{j}{k}\gamma_j, \qquad k = 0, \ldots, d \tag{3}$$

The "equations" (3) can clearly be "solved," i.e., the γ_j's can be expressed by the f_k's. This shows that although the definition of the numbers γ_k apparently depends on the choice of the vector w, actually

the numbers γ_k are independent of w. $\qquad$ (4)

Replacing w by -w, it then follows from (1) and (4) that

$$\gamma_k = \gamma_{d-k}, \qquad k = 0, \ldots, d \tag{5}$$

Combining (3) and (5) we obtain

$$f_k = \sum_{j=0}^{[d/2]} \left[\binom{j}{k} + (1 - \delta(d,2j))\binom{d-j}{k} \right]\gamma_j, \qquad k = 0, \ldots, d \tag{6}$$

It now remains to evaluate γ_j. In order to do so, we introduce a *k-incidence*, k = 0,...,d - 1, as a pair (X,x) where X is a facet of Q and x is a vertex of X such that the number of edges in X containing x and being oriented towards x is k. (It is understood that we have made a choice of the vector w.) Denoting by I_k the number of k-incidences in Q, we then have

$$I_k \leq v \cdot \gamma_k, \qquad k = 0, \ldots, d - 1 \tag{7}$$

where v denotes the number of facets of Q. In fact, each facet X of Q is itself a simple polytope, cf. (c), and therefore we have numbers γ_k^X, k = 0,...,d - 1, associated with X in the same way as we have numbers γ_k associated with Q. It is clear that a vertex x of X contributes to γ_k^X if and only if (X,x) is a k-incidence. On the other hand, by choosing the vector w such that each vertex of Q not in X is "below" every vertex of X, we see that $\gamma_k^X \leq \gamma_k$. Since $I_k = \sum \gamma_k^X$, where we sum over all facets X of Q, we see that (7) holds.

We also have

$$I_k = (d-k)\gamma_k + (k+1)\gamma_{k+1}, \qquad k = 0,\ldots,d-1 \qquad (8)$$

In fact, for each vertex x of Q there are d facets containing x, there are d edges containing x, and each facet contains d - 1 of the d edges, cf. (a). Therefore, a vertex x contributes to I_k if and only if either x has in-degree k, in which case the contribution is d - k, or x has in-degree k + 1, in which case the contribution is k + 1. This proves (8).

Now combining (7) and (8) one easily shows that

$$\gamma_k \leq \binom{v-d+k-1}{k}, \qquad k = 0,\ldots,d \qquad (9)$$

Taking

$$\phi_k(v,d) = \sum_{j=0}^{[d/2]} \left[\binom{j}{k} + (1 - \delta(d,2j))\binom{d-j}{k} \right] \binom{v-d+j-1}{j}$$

$$k = 0,\ldots,d-2$$

we then see by (6) that (*) holds.

To prove that (**) holds when Q is the dual of a neighborly polytope we shall prove that

$$\gamma_k = \binom{v-d+k-1}{k}, \qquad k = 0,\ldots,[d/2], \text{ when Q is the dual} \qquad (10)$$
of a neighborly polytope.

To see this, suppose that we have strict inequality in (7) for some $k \leq [d/2] - 1$. Then there is a facet X of Q such that $\gamma_k^X < \gamma_k$, whence some vertex x of Q not in X has in-degree k. The out-degree therefore is d - k, cf. (1). Let G be the (d - k)-face of Q determined by this (d - k)-out-star, cf. (b). Supposing that w has been chosen in such a way that each vertex of Q not in X is "below" each vertex of X, it follows that $G \cap X = \emptyset$, and since G is the intersection of k facets $X_1,\ldots,X_k$, we see that the k + 1 facets $X, X_1, \ldots, X_k$ have an empty intersection. When Q is the dual of a neighborly polytope this is impossible by (d). This shows that for $k = 0,\ldots,[d/2] - 1$ we have equality in (7) when Q is the dual of a neighborly polytope. This implies (10), and so the entire proof is completed.

REFERENCES

1. Aa. Bondesen and A. Brøndsted. A dual proof of the upper bound conjec-
    ture for convex polytopes, *Math. Scand.* 46(1980), 95-102.

2. P. McMullen. The maximum numbers of faces of a convex polytope, *Mathe-
    matika* 17(1970), 179-184.

EULER'S RELATION AND WHERE IT LED

G. Thomas Sallee

Department of Mathematics
University of California at Davis
Davis, California

1. INTRODUCTION

Let P^d denote the class of all convex polytopes in E^d, d-dimensional Euclidean space. It is well-known that if P is a d-dimensional polytope, it satisfies Euler's relation

$$\sum_{i=-1}^{d} (-1)^i f_i(P) = 0 \tag{1.1}$$

where $f_i(P)$ denotes the number of i-dimensional faces of P, counting $\emptyset$, the empty set, to be the sole face of dimension -1 (see Grünbaum [6]; the reader may consult this source for other standard results and definitions). It will also be convenient for us to write Euler's relation as:

$$\sum (-1)^{\dim F} = 1 \tag{1.2}$$

where the summation is taken over all nonempty faces of a polytope P.

In the past dozen years, several papers [7, 14-16, 19, 20, 24, 26, 27, 32-34] have studied functions which satisfy *Euler-type relations*; that is, functions which have the property that

45

$$\sum (-1)^{\dim F} \varphi(F) = \varepsilon\varphi(P), \tag{1.3}$$

again where the sum is taken over all nonempty faces of P. Because of the frequency with which we will use the expression on the left-and side of (1.3), we will call it the *derived function* of φ and denote it as $\varphi^*(P)$. An Euler-type relation then takes the form $\varphi^*(P) = \varepsilon\varphi(P)$.

The most important results in discussing Euler-type relations have to do with their connection to valuations. A *valuation* ψ mapping P^d to a vector space V satisfies the equality

$$\psi(P \cup Q) + \psi(P \cap Q) = \psi(P) + \psi(Q) \tag{1.4}$$

whenever P $\cup$ Q is convex. (It suffices to establish (1.4) when P $\cap$ Q is a face of each [24].)

Here we wish to explore the connection between the two concepts, prove some of the many corollaries of the Euler-type relation, and to call attention to the possibilities of proving many of these results by means of incidence algebras.

2. VALUATIONS

Valuations seem to appear at every turn in a geometric context and most of the functions of interest are, in fact, valuations. That this occurs is not too surprising, for the most basic functions in geometry--characteristic functions and support functions--are valuations. (The former fact is obvious, while the latter is proved in [24].) Moreover, it is clear that if ψ is a valuation, integrating ψ with respect to any measure will simply produce a new valuation.

Hence, simply by checking the definitions, because the following functions arise by integrating an appropriate characteristic function with respect to an appropriate measure, they are valuations:

(2.1) $G_0(P)$, the number of lattice points inside a polytope [2];

(2.2) $\beta(F,P)$, the interior angle of F in the polytope P ([6], p. 297);

(2.3) $\gamma(F,P)$, the exterior angle of F in the polytope P ([6], p. 308);

(2.4) $\gamma^{m,d}(C^k)$, the Grassman angles of the cone C [7];

(2.5) $\delta(F,P)$, the angle deficiency of P at F [32]; and

(2.6) $Z(P)$, the moment vector of P [26].

Recall that the *support function*, H(P,u) equals sup{<p,u>; p ∈ P,u ∈ S^{d-1}}, where < , > denotes inner product. Beginning with the support function, it is clear that the following functions are valuations:

    (2.7)   m(P), the mean width of P [33];

    (2.8)   S(P), the Steiner point of P [31];

    (2.9)   mixed volumes. This last demonstration is rather more involved and carried out by Shephard in [34].

    (2.10) $W_i(K)$, the quermassintegrals [8].

In addition, Shephard [34] remarks that we get different valuations if H(P,u) is replaced by g(P,u) = sup{f(<x,u>) : x ∈ P,u ∈ S^{d-1}} for any monotone function f.

Finally, there is one more consequence of the fact that H(P,u) is a valuation of a rather different flavor:

    (2.11) P + Q = (P ∪ Q) + (P ∩ Q) whenever P,Q and P ∪ Q are convex [15].

What is the connection between all of these functions and Euler-type relations? There are two principal results. The first appears in [24], while the second occurs with a somewhat different emphasis in [15], Theorems 1 and 15. An example is also given in [24] showing that continuity is an essential hypothesis to the first result.

(2.12) THEOREM. Suppose φ is a function, continuous with respect to the Hausdorff metric, which satisfies an Euler-type relation involving faces of more than one dimension. Then φ is a valuation.

(2.13) THEOREM. Suppose φ is a valuation on P^d with the property that φ(P + t) = φ(P) + ψ(P,t) where ψ is linear in t. Then φ = $φ_0$ + $φ_1$ + ··· + $φ_d$ where $φ_k(F^j)$ = 0 if j < k, $φ_k(nP)$ = $n^k φ(P)$ and $φ_k^*(P)$ = $(-1)^k φ_k(-P)$.

Theorem (2.13) now allows us to conclude that the support function satisfies an Euler-type relation. For H(P + t,u) = H(P,u) + <t,u> for any fixed u, and H(nP,u) = H(P,u). Thus, by (2.13), H*(P,u) = (-1)H(-P,u).

There has been extensive work in the past few years with efforts to characterize certain types of valuations.

Those interested in questions of this type should consult [8, 10, 17, 18, 25, 37]. In much the same spirit are the questions involving

Minkowski-additive functions (ν is such a function if ν(A + B) = ν(A) + ν(B)). Because of (2.11), it is clear that such functions are valuations, but the converse is clearly false. See [18, 28-30, 36] for results to date in this direction. Finally, efforts have been made to extend valuations from P^d to unions of cell complexes. See, in particular, [4, 5, 11, 13, 19].

3. CONSEQUENCES OF EULER'S RELATION

One might infer from the preceding section that only valuations satisfy Euler-type relations. This is not true. We can begin with any function and find associated functions satisfying Euler-type relations. The following result was originally proved for valuations in [24], but M. A. Perles remarked that the proof was valid in the general case. Notice how it dovetails with McMullen's decomposition theorem for valuations (2.13).

(3.1) THEOREM. Suppose φ : P^d → V is a vector space. Then φ = λ + ν where λ and ν each satisfy an Euler-type relation with λ*(P) = λ(P) and ν*(P) = -ν(P).

The proof of this result is a simple corollary of

(3.2) LEMMA. Suppose φ* is the derived function of φ. Then (φ*)* = φ.

From (3.2), we can prove (3.1) by setting λ = ½(φ + φ*) and ν = ½(φ − φ*).

The proof of (3.2) will be deferred to the next section where it will be shown to be a corollary of Euler's relation.

Euler-type relations are now rather common in the literature. The first, besides Euler's relation itself, is the Gram-Euler relation [3]

$$\sum (-1)^{\dim F} \beta(F,P) = 0 \tag{3.3}$$

where β(F,P) is the interior angle of F in P. Since, Grassman angles [7], Steiner points [31], mean widths [33], mixed volumes [34], and a host of other functions have been shown to satisfy Euler-type relations.

Once a function satisfies an Euler-type relation, it will also satisfy further identities. Recall, for example, the Dehn-Sommerville Equations. These were originally stated for simplicial polytopes [35], but the more natural generalization [7], would seem to be the *simple polytopes*--that is, those d-polytopes for which each vertex lies in exactly d facets.

(3.4) (Dehn-Sommerville Equations). If P is a simple d-polytope then

$$\sum_{j=0}^{i} (-1)^j \binom{d - j}{d - i} f_j(P) = f_i(P) \quad \text{for} \quad i = 0,\ldots,d \tag{3.5}$$

Note that the case $i = d$ is just Euler's relation. Perles has shown that only $[d/2]$ of these equations are linearly independent ([6], p. 146).

A similar set of equations hold for arbitrary functions satisfying an Euler-type relation [15].

(3.5) THEOREM. Let φ be a function mapping P^d to V, a vector space, such that $\varphi*(P) = \varepsilon\varphi(P)$. Then for any simple polytope, S,

$$\sum_{j=0}^{i} (-1)^j \binom{d - j}{d - i} \sum \varphi(F^j) = \varepsilon \sum \varphi(F^i) \tag{3.6}$$

where the summations are taken over all faces of S of appropriate dimension.

Proof. Let F_0^i -e any i-face of S. Then since φ satisfies an Euler-type relation,

$$\sum_{j=0}^{i} (-1)^j \sum_{F^j \subseteq F_0^i} \varphi(F^j) = \varepsilon\varphi(F^i)$$

Sum both sides of this equation over all i-faces of S. Since S is simple, each j-face is determined by exactly $d - j$ facets and each set of $d - i$ of them determine an i-face. Thus, each j-face is counted $\binom{d - j}{d - i}$ times and the proof is complete. □

Special cases of the result above have been known for some time [7, 20, 32].

4. INCIDENCE ALGEBRAS

Now we wish to present an efficient framework for dealing with the types of problems which we have been considering. It relies on the notion of an incidence algebra, due to G. C. Rota [21]. While all of the details are in the original paper, a brief summary is in order.

Let $L(P)$ be the lattice of faces of a d-polytope, including the empty face, $\emptyset$, and P. In the lattice, $F \leq G$ if and only if $F \subseteq G$. The *incidence algebra* $A(P) = \{\sigma : \sigma$ maps $L \times L$ to a ring R, $\sigma(F,G) = 0$ if $F \nsubseteq G\}$. Addition is defined in the obvious way and multiplication by $\sigma \circ \tau(F_1, F_2) = \sum \sigma(F_1, G)\tau(G, F_2)$, where the summand vanishes unless $F_1 \subseteq G \subseteq F_2$.

Three functions are of particular interest:

$$\iota(F,G) = \begin{cases} 1, & \text{if} \quad F = G \\ 0, & \text{otherwise} \end{cases}$$

$$\zeta(F,G) = \begin{cases} 1, & \text{if} \quad F \subseteq G \\ 0, & \text{otherwise} \end{cases}$$

$\nu(F,G) = $ number of faces H such that $F \subseteq H \subseteq G$

In addition, we adopt the convention for any $\omega \in A$ that $\bar{\omega}(F,G) = (-1)^{\dim G - \dim F} \omega(F,G)$. Euler's relation may now be stated in the very compact form

$$\zeta \circ \bar{\zeta} = \iota = \bar{\zeta} \circ \zeta \tag{4.1}$$

To see the equivalence of (4.1) to Euler's relation, observe that for any two faces $F_1 \subseteq F_2$,

$$\zeta \circ \bar{\zeta}(F_1, F_2) = \sum \zeta(F_1, F)\bar{\zeta}(F, F_2)$$
$$= \sum (-1)^{\dim F_2 - \dim F} = (-1)^{\dim F_2} \sum (-1)^{\dim F}$$

where the sum is taken over all faces F such that $F_1 \subseteq F \subseteq F_2$.

If $\zeta \circ \bar{\zeta} = \iota$, and $F_1 \neq F_2$ this last sum must equal 0, which is the strong form of Euler's relation ([7], p. 137). Conversely, this form of Euler's relation implies the sum is 0.

The other equality may be proved in a similar fashion.

We wish to work with (4.1) in the content of the algebra and see what corollaries can be drawn. Three corollaries are immediate.

$$\bar{\zeta} \circ \zeta \circ \zeta = \zeta \tag{4.2}$$

$$\zeta \circ \zeta \circ \bar{\zeta} = \zeta \tag{4.3}$$

$$\zeta \circ \zeta \circ \bar{\zeta} \circ \bar{\zeta} = \iota \tag{4.4}$$

Using the easily-provable facts that

$$\zeta \circ \zeta = \nu \tag{4.5}$$

and

$$\bar{\zeta} \circ \bar{\zeta} = \bar{\nu} \tag{4.6}$$

the relations above can be translated into more familiar terms.

$$\sum (-1)^{\dim F} \nu(F, F_2) = (-1)^{\dim F_1} \tag{4.7}$$

$$\sum (-1)^{\dim FF} \nu(F_1, F) = (-1)^{\dim F_2} \tag{4.8}$$

$$\sum (-1)^{\dim F} \nu(F_1, F) \nu(F, F_2) = 0 \qquad \text{if} \quad F_1 \neq F_2 \tag{4.9}$$

In each case the summation is taken over all faces F such that $F_1 \subseteq F \subseteq F_2$.

Just as Euler's relation is essentially the only linear relation satisfied for the numbers of faces of every convex polytope, (4.7), (4.8) and (4.9) are also the only relations of their form holding for every convex polytope.

We may also employ incidence algebras to deduce similar kinds of relations for any function satisfying an Euler-type relation. To see this, let φ be an arbitrary function mapping P^d into our ring R, with $\varphi(\emptyset) = 0$. Define two auxiliary functions in the algebra: $\varphi_1(F, G) = \varphi(F)$ if $F \subseteq G$; $\varphi_2(F, G) = \varphi(G)$ if $F \subseteq G$. (Note that φ_1 and φ_2 depend on the polytope P in which F and G lie.) Also let $\alpha(P) = (-1)^{\dim P}$. Then

$$\alpha_1 \circ \alpha_1 = \iota = \alpha_2 \circ \alpha_2 \tag{4.10}$$

More important:

$$\varphi_2 \circ \alpha_1(\emptyset, P) = \varphi_2^*(\emptyset, P) \tag{4.11}$$

To see this, by definition,

$$\varphi_2 \circ \alpha_1(\emptyset, P) = \sum \varphi_2(\emptyset, F)\alpha_1(F, P) = \sum \varphi(F)(-1)^{\dim F}$$
$$= \varphi^*(P) = \varphi_2^*(\emptyset, P)$$

From (4.10) and (4.11), we may complete the deferred proof of

$$(\varphi^*)^*(P) = \varphi(P) \tag{3.2}$$

For $(\varphi^*)^*(P) = (\varphi_2^*)(\emptyset,P) = \varphi_2^* \circ \alpha_1(\emptyset,P) = \varphi_2 \circ \alpha_1 \circ \alpha_1(\emptyset,P)$

$\qquad\qquad = \varphi_2(\emptyset,P) = \varphi(P)$

Note that we have produced a virtually painless proof of a central result in this area by a simple application of Euler's relation.

In a similar way, we may extend the reasoning behind (4.7), (4.8) and (4.9) to arbitrary functions φ to get results of the following type:

(4.12) PROPOSITION. If φ is a function on the faces of P to a ring R, with $\varphi(\emptyset) = 0$, and $\Phi(F) = \sum \varphi(G)$, then $\varphi(F) = \sum (-1)^{\dim G}\Phi(G)$.

The preceding result is very reminiscent of the Möbius inversion formula and is, in fact, just the geometric analogue of the formula.

Both of the last two results are examples of the type of inversion formulas which McMullen has applied to successfully. In [15], he proved Ehrhart's "reciprocity law" [2] on the number of lattice points inside a rational polytope, and in [14] he extended the technique to other relations.

The idea is that indicated above except that we may begin with any pair of inverse functions σ,τ.

(4.13). Let φ be a function on the faces of P, and suppose $\sigma,\tau \in A(P)$ such that $\sigma \circ \tau = \iota$. If $\psi(G) = \sum \sigma(\emptyset,F)\varphi(F)$, then $\varphi(G) = \sum \tau(\emptyset,F)\psi(F)$, where each sum is taken over all faces $F \subseteq G$.

While the result above is completely general (and proved in the same way as the special cases), it loses its appeal unless φ and τ are of intrinsic interest. McMullen has generally done his inversions using the relation:

$$\beta \circ \bar{\gamma} = \iota = \bar{\beta} \circ \gamma \qquad\qquad (4.14)$$

where $\beta(F,G)$ and $\gamma(F,G)$ are the interior and exterior angles of F in G, respectively. His proof, in turn, is based on an old theorem of Sommerville which states that for a pointed cone K with vertex zero,

$$\sum (-1)^{\dim F}\beta(F,K) = (-1)^{\dim K}\beta(0,K) \qquad\qquad (4.15)$$

But even this follows from Euler's relation as can be seen by a close examination of the proof of the analogous statement for polytopes given in Grünbaum ([7], p. 298).

5. UNSOLVED PROBLEMS AND COMMENTS

1. In [17] McMullen makes the following conjecture, The family of continuous, translation-invariant valuations on K^d (compact convex sets in E^d) is just the (weak) completion of the linear space of mixed volumes.

2. Valuations are characterized by the values which they take on a relatively small collection of sets. Hadwiger [9], Berg [1], Meier [18] use collections large enough to fully determine the valuations, but no attempt seems to have been made to determine a minimal set. So suppose G is a group of motions of E^d, and ψ is a valuation invariant under G. Find a collection on each member of C, which then determines ψ on all of P^d (or K^d).

3. Are there any more geometrically interesting, basic inverses than already known? There are only a few known to date:

$$\zeta \circ \bar{\zeta} = \iota$$

$$\nu \circ \bar{\nu} = \iota$$

$$\beta \circ \bar{\gamma} = \gamma \circ \bar{\beta} = \iota$$

The first is Euler's relation; the second follows from it and the third may follow. Are there more? Note that an inverse can be found for almost any function $\sigma \in A(P)$, but in general it has no intuitive appeal.

4. If we begin with a valuation φ on P^d and form the function

$$\psi(G) = \sum \sigma(F,G)\varphi(F)$$

under what conditions is ψ a valuation? In [24] it is shown that when $\sigma(F,G) = (-1)^{\dim F}$, ψ is a valuation.

REFERENCES

1. C. Berg. Abstract Steiner points for convex polytopes, *J. London Math. Soc.* (2)4(1971), 176–180.

2. E. Ehrhart. Démonstration de la loi de réciprocité pour un polyèdre entier, *C. R. Acad. Sci. Paris* 265A (1967), 5–7.

3. J. P. Gram. Om rumvinklerne i et Polyeder, *Tidsskr. Math.* (Copenhagen)
    (3)4(1874), 161-163.

4. H. Groemer. Eulersche Charakteristik, Projektionen und Quermassintegrale,
    Math. Ann. 198(1972), 23-56.

5. H. Groemer. On the extension of additive functionals on classes of convex
    sets, *Pac. J. Math.* 75(1978), 397-410.

6. B. Grünbaum. *Convex Polytopes*, Wiley, New York, 1967.

7. B. Grünbaum. Grassmann angles of convex polytopes, *Acta Math.* 121(1968),
    293-302.

8. H. Hadwiger. Translationsinvariante, additive und schwachstetige Polyeder-
    funktionale, *Arch. Math.* 3(1952), 387-394.

9. H. Hadwiger. *Vorlesungen über Inhalt, Oberfläche und Isoperimetrie,*
    Springer, Berlin-Göttingen-Heidelberg, 1957.

10. H. Hadwiger. Homothetienvariante und additive Polyderfunktionen, *Arch.
    Math.* 25(1974), 203-205.

11. V. Klee. The Euler characteristic in combinatorial geometry, *Amer. Math.
    Monthly* 70(1963), 119-127.

12. I. G. MacDonald. Polynomials associated with finite cell complexes, *J.
    London Math. Soc.* (2)4(1971), 181-192.

13. P. Mani. On angle sums and Steiner points of polyhedra, *Israel J. Math.*
    9(1971), 380-388.

14. P. McMullen. Non-linear angle-sum relations for polyhedral cones and
    polytopes, *Math. Proc. Comb. Phil. Soc.* 78(1975), 247-261.

15. P. McMullen. Valuations and Euler-type relations on certain classes of
    convex polytopes, *Proc. London Math. Soc.* (3)35(1977), no. 1, 113-135.

16. P. McMullen. Lattice invariant valuations on rational polytopes, *Arch.
    Math.* 31(1978), 509-516.

17. P. McMullen. Continuous translation invariant valuations on the space of
    compact convex sets, *Arch. Math.* to be published.

18. Ch. Meier. Multilinearität bei Polydederaddition, *Arch. Math.* 29(1977),
    210-217.

19. M. A. Perles and G. T. Sallee. Cell complexes, valuations and the Euler
    relation, *Canad. J. Math.* 22(1970), 235-241.

20. M. A. Perles and G. C. Shephard. Angle sums of convex polytopes, *Math.
    Scand.* 21(1967), 199-218.

21. G. C. Rota. On the foundations of combinatorial theory. I. Theory of
    Möbius functions, *Z. Wahrscheinlichkeitstheorie* 2(1964), 340-368.

22. G. C. Rota. On the combinatorics of the Euler characteristic. *Studies
    in Pure Mathematics* (Presented to Richard Rado), Academic Press, London,
    1971, 221-233.

23. G. T. Sallee. A valuation property of Steiner points, *Mathematika* 13
    (1966), 76-82.

24. G. T. Sallee. Polytopes, valuations and the Euler relation, *Canad. J.
    Math.* 20(1968), 1412-1424.

25. R. Schneider. On Steiner points of convex bodies, *Israel J. Math.* 9(1971), 241-249.

26. R. Schneider. Krümmungsschwerpunkte konvexer Körper, *Abh. Math. Sem. Univ. Hamburg* (I) 37(1972), 112-132.

27. R. Schneider. (II) 37(1972), 204-217.

28. R. Schneider. Bewegungsäquivariante, additive und stetige Transformationen konvexer Bereiche, *Arch. Math.* 25(1974), 303-312.

29. R. Schneider. Equivariant endomorphisms of the space of convex bodies, *Trans. Amer. Math. Soc.* 194(1974), 53-78.

30. R. Schneider. Additive transformationen konvexer Körper, *Geom. Ded.* 3 (1974), 221-228.

31. G. C. Shephard. The Steiner point of a convex polytope, *Canad. J. Math.* 18(1966), 1294-1300.

32. G. C. Shephard. Angle deficiencies of convex polytopes, *J. London Math. Soc.* 43(1968), 325-336.

33. G. C. Shephard. The mean width of a convex polytope, *J. London Math. Soc.* (1)43(1968), 207-209: MR 37#2087.

34. G. C. Shephard. Euler-type relations for convex polytopes, *Proc. London Math. Soc.* (3)18(1968), 597-606: MR 38#606.

35. D. M. Y. Sommerville. The relations connecting the anglesums and volume of a polytope in space of n dimensions, *Proc. Royal Soc. London*, Ser. A, 115(1927), 103-119.

36. W. Spiegel. Zur Minkowski-additivität bestimmter Eikörperabbildungen, *J. Reine Angew. Math.* 286(1976), 164-168.

37. W. Spiegel. Ein Beetrag über additive, translationsinvariante stetige Eikörperfunktionale, *Geom. Ded.* 7(1978), 9-19.

MIXED VOLUMES AND GEOMETRIC INEQUALITIES

G. D. Chakerian

Department of Mathematics
University of California at Davis
Davis, California

The objective here is to pursue a discussion of how the additivity and monotonicity of Minkowski's mixed volumes can be used to obtain relationships between geometric invariants of sets of constant width and estimates on area and volume.

1. PROPERTIES OF MIXED VOLUMES

We shall consider convex bodies in d-dimensional Euclidean space R^d. The *vector sum*, or *Minkowski sum*, of convex bodies K_1 and K_2 is

$$K_1 + K_2 = \{x + y : x \in K_1 \text{ and } y \in K_2\} \tag{1}$$

If K is a convex body and $\lambda \geq 0$, then λK is again a convex body, where

$$\lambda K = \{\lambda x : x \in K\} \tag{2}$$

The *outer parallel set* of K at distance $\lambda \geq 0$ is $K + \lambda B$, where B is the unit ball centered at the origin in R^d. The volume of $K + \lambda B$ is a polynomial of degree d in λ,

57

$$V(K + \lambda B) = \sum_{k=0}^{d} \binom{d}{k} W_k(K) \lambda^k \tag{3}$$

where $W_k(K)$ is the k-th Quermassintegral of K.

More generally, if $K_1, \ldots, K_n$ are convex bodies in R^d and $\lambda_1, \ldots, \lambda_n \geq 0$, then the volume of $\lambda_1 K_1 + \cdots + \lambda_n K_n$ is a homogeneous polynomial in $\lambda_1, \ldots, \lambda_n$,

$$V(\lambda_1 K_1 + \cdots + \lambda_n K_n) = \sum V(K_{i_1}, \ldots, K_{i_d}) \lambda_{i_1} \cdots \lambda_{i_d} \tag{4}$$

where the coefficients $V(K_{i_1}, \ldots, K_{i_d})$ are the *mixed volumes* of Minkowski.

The invention of mixed volumes can be traced back to an interesting letter from Minkowski to David Hilbert dated Dec. 10, 1900 (see [7]). The mixed volume $V(K_1, \ldots, K_d)$ is nonnegative and symmetric in its arguments. Furthermore, if we fix $K_2, \ldots, K_d$ and let $K = K_1$ vary, then the functional f defined by

$$f(K) = V(K, K_2, \ldots, K_d)$$

enjoys the following properties (see Bonnesen and Fenchel [2]):

$$f(K + K') = f(K) + f(K') \qquad \text{(additivity)} \tag{5a}$$

$$\lambda \geq 0 \Rightarrow f(\lambda K) = \lambda f(K) \qquad \text{(homogeneity)} \tag{5b}$$

$$K \subset K' \Rightarrow f(K) \leq f(K') \qquad \text{(monotonicity)} \tag{5c}$$

$$f(x + K) = f(K) \qquad \text{(translation invariance)} \tag{5d}$$

The Quermassintegrals appearing in (3) are just

$$W_k(K) = V(\underbrace{K, \ldots, K}_{d-k}, \underbrace{B, \ldots, B}_{k})$$

2. SETS OF CONSTANT WIDTH

In what follows we shall let K^* denote the reflection of K through the origin, so

$$K^* = \{-x : x \in K\}$$

A convex body K has *constant width* $\lambda > 0$ if

$$K + K^* = \lambda B$$

Barbier's Theorem asserts that all sets of constant width λ in R^2 have the same perimeter $\pi\lambda$. In R^3 we let $M(K) = 3V(K,B,B) = $ the total mean curvature, $S(K) = 3V(K,K,B) = $ the surface area, and $V(K) = V(K,K,K) = $ the volume of K. Then we have two independent relations for sets of constant width λ:

$$M(K) = 2\pi\lambda$$

$$2V(K) = \lambda S(K) - \frac{2\pi}{3}\lambda^3$$

The second of these was first observed by Blaschke [1]. Various authors [4, 5, 6, 8] have generalized this to R^d by showing that the Quermassintegrals of a set of constant width satisfy $\left[\dfrac{d+1}{2}\right]$ independent relations. In case K has constant width 1 these relations take the form

$$W_{d-n}(K) = \sum_{k=0}^{n} (-1)^k \binom{n}{k} W_{d-k}(K) \tag{6}$$

It is of interest to note that relationships between mixed volumes generalizing (6) can be derived using essentially only the symmetry and additivity properties. First observe that because of additivity the relation $X = Y + Z$ implies

$$V(\ldots,Y,\ldots) = V(\ldots,X,\ldots) - V(\ldots,Z,\ldots)$$

where the undesignated arguments are the same in each term. It follows by induction that if $X_i = Y_i + Z_i$, $i = 1,\ldots,n \leq d$, then $V(Y_1,\ldots,Y_n,*)$ can be expressed as a sum of terms of the form $(-1)^\sigma V(X_{i_1},\ldots,X_{i_k},Z_{i_{k+1}},\ldots,Z_{i_n},*)$, where $*$ represents the same $d - n$ remaining arguments in each term. The signs $(-1)^\sigma$ and permutations $X_{i_1},\ldots,X_{i_k},Z_{i_{k+1}},\ldots,Z_{i_n}$ are obtained by formally expanding the product $(x_1 - z_1)(x_2 - z_2)\ldots(x_n - z_n)$ and then replacing terms of the form $(-1)^\sigma x_{i_1}\ldots x_{i_k} z_{i_{k+1}}\ldots z_{i_n}$ in the expansion by

$$(-1)^\sigma V(X_{i_1},\ldots,X_{i_k},Z_{i_{k+1}},\ldots,Z_{i_n},*)$$

For example, if $X = Y + Z$ we can express $V(Y,\ldots,Y,*)$ in terms of mixed volumes involving Y and Z by formally expanding $(x - z)^n = \sum_{k=0}^{n} (-1)^k \binom{n}{k} x^{n-k} z^k$. Thus

$$V(\underbrace{Y,\ldots,Y}_{n},*) = \sum_{k=0}^{n} (-1)^{k} \binom{n}{k} V(\underbrace{X,\ldots,X}_{n-k},\underbrace{Z,\ldots,Z}_{k},*) \tag{7}$$

If K is of constant width 1 in R^{d}, so $K + K^{*} = B$, we may take $Y = K^{*}$, $Z = K$, $X = B$, and the remaining arguments all equal to B in (7). Using the fact that $W_{k}(K^{*}) = W_{k}(K)$, we then have (6).

A natural generalization of (6), where K is not necessarily of constant width, is obtained as follows. Let $K + K^{*} = E = $ the difference set of K. In (7), take $Y = K^{*}$, $Z = K$, $X = E$, and all the remaining arguments equal to E. We have $V(K^{*},\ldots,K^{*},E,\ldots,E) = V(K,\ldots,K,E,\ldots,E)$ since E is centrally symmetric, and (7) yields

$$V(\underbrace{K,\ldots,K}_{n},\underbrace{E,\ldots,E}_{d-n}) = \sum_{k=0}^{n} (-1)^{k} \binom{n}{k} V(\underbrace{K,\ldots,K}_{k},\underbrace{E,\ldots,E}_{d-k}) \tag{8}$$

In case $n = d = 3$, we obtain from (7)

$$V(Y) + V(Z) = V(X) - 3V(X,X,Z) + 3V(X,Z,Z) \tag{9}$$

Interchanging Y and Z in (9), adding the resulting equations, and using the fact that $Y + Z = X$, we have

$$2(V(Y) + V(Z)) = 3V(X,Y,Y) + 3V(X,Z,Z) - V(X) \tag{10}$$

Equation (10) may be viewed as a relationship between the mixed volumes of Y and Z that must hold if X admits the Minkowski decomposition $X = Y + Z$. In case $X = B$, we have

$$2(V(Y) + V(Z)) = S(Y) + S(Z) - \frac{4\pi}{3} \tag{11}$$

one of several relations between geometric invariants of decompositions of the ball first discovered by W. J. Firey. If $Z = Y*$, (11) reduces to Blaschke's relation for sets of constant width.

As a final example, consider sets K_{1}, K_{2}, K_{3} of constant width 1 in R^{3}. Since $K_{i} + K_{i}^{*} = B$, $i = 1, 2, 3$, we have

$$V(K_{1}^{*}, K_{2}^{*}, K_{3}^{*}) = V(B,B,B) - \sum V(K_{i},B,B) + \sum V(K_{i},K_{j},B) - V(K_{1},K_{2},K_{3}) \tag{12}$$

Since $V(K_1^*, K_2^*, K_3^*) = V(K_1, K_2, K_3)$ and $V(K_i, B, B) = V(B)/2 = 2\pi/3$, $i = 1, 2, 3$, from (12) we obtain

$$2V(K_1, K_2, K_3) = V(K_1, K_2, B) + V(K_2, K_3, B) + V(K_3, K_1, B) - \frac{2\pi}{3} \tag{13}$$

Blaschke's relation is the special case where $K_1 = K_2 = K_3$.

3. ESTIMATES ON VOLUME

The Blaschke–Lebesgue Theorem states that the Reuleaux triangle has the least area of all plane convex sets of the same constant width. The additivity and monotonicity of mixed volumes gives the area of the Reuleaux triangle as a lower bound neatly as follows.

   If K is a plane convex body of constant width 1, we have $K + K^* = B$ so

$$\pi = V(B, B) = V(K + K^*, K + K^*) = 2V(K) + 2V(K, K^*) \tag{14}$$

From a theorem of Pál we know that K admits a circumscribed regular hexagon H. This hexagon has area $\sqrt{3}/2$. Since a translate of K^* also fits inside H, the monotonicity of mixed volumes gives $V(K, K^*) \leq V(H, H) = V(H) = \sqrt{3}/2$. Using this in (14) we obtain

$$V(K) \geq \frac{\pi - \sqrt{3}}{2} \tag{15}$$

The right-hand side of (15) is the area of a Reuleaux triangle of constant width 1.

   Weissbach [9] was able to generalize this kind of argument to obtain estimates on areas of rotors of various types. His proof that any rotor in an equilateral triangle has area at least that of the 60° biangle is as follows. If K is a rotor in an equilateral triangle of height 1, then $K_1 + K_2 + K_3 = B$, where K_1 is a translate of K, and K_{i+1} is obtained from K_i by a positive 120° rotation. The functional properties of mixed volumes lead then to

$$\pi = V(B, B) = V(K_1 + K_2 + K_3, K_1 + K_2 + K_3) = 3V(K) + 6V(K, K') \tag{16}$$

where K' is a rotation of K through 120°. In contrast to the situation for sets of constant width, there is no "suitable" polygon of fixed shape covering all rotors in equilateral triangles. However, by a clever continuity

argument, Weissbach is able to prove the existence of hexagons H and H' cover-
ing K and K' respectively, with $V(H,H') \leq \sqrt{3}/4$. Thus, $V(K,K') \leq V(H,H') \leq \sqrt{3}/4$, so (16) gives

$$V(K) \geq \frac{\pi}{3} - \frac{\sqrt{3}}{2} \tag{17}$$

The right-hand side is the area of a 60° biangle that is a rotor in an equilateral triangle of height 1.

As noted by Weissbach, it is unlikely that arguments such as these can
be generalized to higher dimensions using circumscribed polytopes in a similar
fashion to yield sharp lower bounds on the volumes of sets of constant width.
Note that the Blaschke relation shows that the problem of finding the minimum
volume of a set of constant width 1 in R^3 (still unsolved) is equivalent to
finding the minimum surface area. For a further discussion see [3] and the
references given there.

REFERENCES

1. W. Blaschke. Einige Bemerkungen über Kurven und Flächen von konstanter
    Breite, *Ber. d. Verh. d. Sächs. Akad. Leipzig* 67(1915), 290-297.

2. T. Bonnesen and W. Fenchel. *Theorie der konvexen Körper*, Springer-
    Verlag, Berlin, 1934.

3. G. D. Chakerian. Sets of constant width, *Pacific J. Math.* 19(1966),
    13-21.

4. H. Debrunner. Zu einem massgeometrischen Satz über Körper konstanter
    Breite, *Math. Nach.* 13(1955), 165-167.

5. A. Dinghas. Verallgemeinerung eines Blaschkeschen Satzes über konvexe
    Körper konstanter Breite, *Rev. Math. Un. Interbalkanique* 3(1940), 17-20.

6. H. Groemer. Über die Quermassintegrale von konvexen Körpern konstanter
    Breite, 1966, unpublished manuscript.

7. H. Minkowski. *Briefe an David Hilbert, Mit. Beitr. v. hrsg. von L.
    Rüdenberg u. H. Zassenhaus*, Springer-Verlag, Berlin, 1973.

8. L. A. Santaló. Sobre los cuerpos convexos de anchura constante en E_n,
    Port. Math. 5(1946), 195-201.

9. B. Weissbach. Zur Inhaltsschätzung von Eibereichen, *Wiss. Beitr. Martin-
    Luther-Univ. Halle-Wittenberg M8 Beitr., Algebra Geom.* 6(1977), 27-35.

INTERSECTIONS OF CONVEX SETS AND SURFACES

Paul Goodey*

*Department of Mathematics
University of Oklahoma
Norman, Oklahoma*

If K_1 and K_2 are two planar convex bodies with $K_1 \neq K_2$ and int $K_1 \cap$ int $K_2 \neq \emptyset$ we put

$$\alpha(K_1, K_2) = \# \text{ of connected components of } \partial K_1 \cap \partial K_2$$

Fujiwara [3] and Bol [1] gave the following characterization of discs:

1. K is a circular disc if and only if $\alpha(K, K') = 2$ for all congruent copies K' of K.

More modern proofs of this result yield the following generalization:

2. K_1 and K_2 are congruent discs if and only if $\alpha(K_1, K_2') = 2$ for all congruent copies K_2' of K_2.

In [5] Peterson made some observations concerning $\alpha(K_1, K_2)$ when K_1 and K_2 are sets of constant width. He conjectured that S is of constant width w if and only if $\alpha(S, C)$ is even or infinite for all circles C of diameter w. This conjecture is verified by the following recent result of Goodey and Woodcock:

Current affiliation: Department of Mathematics, Royal Holloway College, London University, London, England

3.   S and C are sets of the same constant width if and only if $\alpha(S,C')$ is even or infinite for all congruent copies C' of C.

The techniques used in the proof of 3 lead one to consider the group of translations rather than the group of all congruences. So it is natural to try to find an analogue of 2 in the case of translations. This analogue was obtained in [4]:

4. K_1 is a translate of K_2 if and only if $\alpha(K_1,K_2') = 2$ for all translates K_2' of K_2.

We note that 4 is a strengthening of 2. For, if $\alpha(K_1,K_2') = 2$ for all congruent copies K_2' of K_2 then 4 implies that K_1 is a translate of every rotation of K_2. Thus, K_2 is a translate of each of its rotations. It follows that K_1 and K_2 are congruent discs.

Yanagihara [6] used 1 to prove the following:

5.   If K is a 3-dimensional convex body such that for any congruent copy K' of K, the boundaries of K and K' intersect in a planar curve (assuming $K \neq K'$ and int K $\cap$ int $K' \neq \emptyset$) then K is a ball.

So it is natural to look for the analogous extension of 4. In this case we obtain the following result:

6.   Let K_1 and K_2 be convex bodies in E^d. If, for every translate K_1' of K_2, $\partial K_1 \cap \partial K_2'$ is the surface of a $(d - 1)$-dimensional convex body (assuming $K_2' \neq K_1$ and int $K_2' \cap$ int $K_1 \neq \emptyset$) then K_1 and K_2 are both translates of the same ellipsoid.

Again it can be seen that this is a strengthening of 5.

It is apparent that from a geometrical point of view the hypothesis in 4 (and 6) is rather too strong. In many ways it would seem more reasonable to work with the hypothesis $\alpha(K_1,K_2') \leq 2$ for all translates K_2' of K_2. In this setting we use some notions from Firey's paper [2]. We say that K_2 *can be translated freely inside* K_1 if for each $\mathbf{x} \in \partial K_1$ there is a translation $\mathbf{t}$ such that $gK_2 = K_2 + \mathbf{t} \subset K_1$ and $\mathbf{x} \in \partial (gK_2) \cap \partial K_1$.

7. K_2 can be translated freely inside K_1 if $\alpha(K_1,K_2') \leq 2$ and $K_1 \setminus K_2' \neq \phi$ for all translates K_2' of K_2.

The analogous result in higher dimensions is:

8.   Let K_1 and K_2 be convex bodies in E^d. If, for every translate K_2' of K_2, $\partial K_1 \cap \partial K_2'$ is empty or the surface of a $(d - 1)$-dimensional convex body (assuming $K_2' \neq K_1$ and int $K_2' \cap$ int $K_1 \neq \emptyset$) then K_1 and K_2 are homothetic ellipsoids.

REFERENCES

1. G. Bol. Zur kinematischen Ordnung ebener Jordan-Kurven, *Abh. Math. Sem. Univ. Hamburg* 11(1936), 394-408.

2. W. Firey. Inner contact measures, *Mathematika* 26(1979), 106-112.

3. M. Fujiwara. Ein Satz über konvexe geschlossene Kurven, *Sci. Repts. Tôhoku Univ.* 9(1920), 289-294.

4. P. R. Goodey and M. M. Woodcock. Intersections of convex bodies with their translates, to be published in *The Geometric Vein-Coxeter Symposium,* Springer-Verlag.

5. B. Peterson. Do self-intersections characterize curves of constant width?, *Amer. Math. Monthly* 79(1972), 505-506.

6. K. Yanagihara. A theorem on surface, *Tôhoku Math. Journal* 8(1915), 42-44.

NONNEGATIVE, MOTION-INVARIANT
VALUATIONS OF CONVEX POLYTOPES

Wolfgang Spiegel

Gesamthochschule Wuppertal Fachbereich 7 (Mathematik)
Wuppertal
Federal Republic of Germany

1. INTRODUCTION

We consider the set of convex polytopes $P(E^d)$ in d-dimensional Euclidean space E^d. A function $f : P(E^d) \to \mathbb{R}$ is called a valuation if the equation

$$f(A \cup B) + f(A \cap B) = f(A) + f(B)$$

holds for all $A, B \in P(E^d)$ with $A \cup B \in P(E^d)$. We say a valuation is motion-invariant if for every rigid motion τ of E^d

$$f(A^{\tau}) = f(A)$$

for all $A \in P(E^d)$, and a valuation is called nonnegative if

$$f(A) \geq 0$$

for all $A \in P(E^d)$.

It is clear that the definitions above have their analogies in the set $K(E^d)$ of all nonvoid, compact, convex sets (convex bodies) of E^d. Hadwiger showed that every motion-invariant valuation which is continuous with respect to the Hausdorff-metric (monotonic with respect to the inclusion of sets resp.) is a linear combination of the "Quermassintegral." It is well-known that in the case of a valuation of $K(E^d)$ one cannot weaken the continuity condition or the condition of monotony, but the question arises of what happens when only those valuations defined on $P(E^d)$ are considered. This question seems reasonable when Hadwiger's proof is studied. In the case of continuous valuations it is necessary that this valuation is defined on $K(E^d)$ ([2], pp. 213, 224, Theorem III). Hadwiger does not make the usual conclusion, that is, verifying the statement for convex polytopes and then going to the limit with arguments of continuity, weakening the condition of continuity, and considering bounded valuations on $P(E^d)$ (that means there exists $M \in \mathbb{R}_+$ and $|f(A)| \leq M$ for all $A \subseteq W$ and W is a cube); one finds an answer in [1] for the two-dimensional case: Every motion-invariant bounded valuation of $P(E^2)$ is a linear combination of the Quermassintegral.

We now consider those motion-invariant valuations on $P(E^d)$ which are nonnegative.

2. THE TWO-DIMENSIONAL CASE

THEOREM 1. Every motion-invariant, nonnegative valuation f of $P(E^2)$ is a linear combination of the Quermassintegral, which means there exists $c_i \in \mathbb{R}$, $0 \leq c_i$ ($i = 0,1,2$) such that

$$f(A) = \sum_{i=0}^{2} c_i W_i(A)$$

holds for all $A \in P(E^2)$. (By $W_i(A)$ we have denoted the Quermassintegral.)

COROLLARY. Considering valuations f of $P(E^2) \cup \{\emptyset\}$ and making the convention $f(\emptyset) = 0$ one obtains: A motion-invariant valuation of $P(E^2) \cup \{\emptyset\}$ is monotonic if and only if this valuation is nonnegative. Monotonic means: $A \subseteq B$ implies $f(A) \leq f(B)$.

Proof. First we consider the real axis E^1 and study the behavior of $f\big|_{P(E^1)} \equiv f$. The set $A \in P(E^1)$ is a closed and bounded nonvoid interval and we can write

$$A = [a,b] = a \times [0,b - a] = a \times (b - a)[0,1] = a \times V(A)[0,1]$$

("$\times$" is the Minkowski-addition: $A \times B = \{x \mid x = a + b, a \in A, b \in B\}$, and $V(A)$ is the one-dimensional volume of A.)

We see that $\psi(A) = f(A) - f(0)$ is simply additive in $P(E^1)$. (A valuation of $P(E^d)\psi$ is called simply additive if $\psi(A) = 0$ for all $A \in P(E^d)$ and dim $A < d$.)

Let α,β be nonnegative real numbers. Since $(\alpha + \beta)[0,1] = \alpha[0,1] \times \beta[0,1] = [0,\alpha] \cup [\alpha,\alpha + \beta]$ and dim$([0,\alpha] \cap [\alpha,\alpha + \beta]) = 0$ we obtain, minding that ψ is invariant under translations:

$$\psi((\alpha + \beta)[0,1]) = \psi(\alpha[0,1]) + \psi(\beta[0,1])$$

Since $\psi(nA) = n\psi(A)$, $n \in \mathbb{N}$, we obtain $f(nA) = f(0) + n(f(A) - f(0))$, $n \in \mathbb{N}$, and thus $f(nA) \geq 0$ implies: $f(A) \geq f(0)$, which means $\psi(A) \geq 0$. So we have $\psi(\alpha[0,1]) \geq 0$ for all $\alpha \in \mathbb{R}_+$ and by the theory of Cauchy's functional equation we obtain $\psi(\alpha[0,1]) = \alpha\psi([0,1])$. Finally we get $\psi(A) = V(A)\psi([0,1])$ and further $f(A) = f(0) + V(A)\psi([0,1])$ and $\psi([0,1]) \geq 0$. Since $f(\lambda A) = f(0) + \lambda V(A)\psi([0,1]) \geq 0$, $\lambda \in \mathbb{R}_+$, we obtain $f(0) \geq 0$. $W_0'(A) = V(A)$, $W_1'(A) = 2$ and $c_1' = \frac{f(0)}{2}$, $c_0' = \psi([0,1])$ give $f(A) = c_1'W_1'(A) + c_0'W_0'(A)$, $c_1', c_0' \geq 0$.

Because of $W_2(A) = \frac{\pi}{2} W_1'(A)$, $W_1(A) = W_0'(A)$ for all $A \in P(E^1)$ we obtain with $c_1 = c_0'$, $c_2 = \frac{2}{\pi} c_1'$

$$f(A) = c_1 W_1(A) + c_2 W_2(A)$$

and this equation holds for all $A \in P(E^2)$ with dim $A < 2$ because f is invariant under rigid motions. We can follow now the argumentation of Hadwiger [2]. The valuation

$$g(A) = f(A) - c_1 W_1(A) - c_2 W_2(A)$$

is simply additive and by a special construction one can find $c_0 \in \mathbb{R}$ with

$$\chi(A) = f(A) - c_0 W_0(A) - c_1 W_1(A) - c_2 W_2(A)$$

is simply additive and fulfills the equation

$$\chi((\alpha + \beta)A) = \chi(\alpha A) + \chi(\beta A) \qquad \alpha, \beta > 0, \qquad A \in P(E^2)$$

We obtain $\chi(\lambda A) = \lambda \chi(A)$ for all $\lambda \in \mathbb{Q}_+$ and since every simply additive, translation-invariant, and rational homogenous valuation is additive in the sense of Minkowski, we obtain $\chi(A \times B) = \chi(A) + \chi(B)$, $A, B \in P(E^2)$. (See [2], p. 63 for further details.)

Let S be a simplex in $P(E^2)$. Since $S \cup S' = A \times B$ with $S' = -S \times t$ and dim $A < 2$, dim $B < 2$, dim $S \cap S' < 2$ we obtain $2\chi(S) = 0$ and so $\chi(A) = 0$ for $A \in P(E^2)$ and the theorem is proved as $f(A) \geqslant 0$ implies $c_0 \geq 0$. $\square$

3. CONSEQUENCES AND OPEN QUESTIONS

First of all let us remark that the additivity in the sense of Minkowski for rational homogeneous, simply additive, translation-invariant valuations holds for all rational homogeneous, translation-invariant valuations of $P(E^d)$. We prove the following generalization of Hadwiger's result mentioned above.

THEOREM 2. Let f be a valuation of $P(E^d)$ which is invariant under translations and for which $f(\lambda A) = \lambda f(A)$ hold for $A \in P(E^d)$ and $\lambda \in \mathbb{Q}_+$. Then we have

$$f(A \times B) = f(A) + f(B)$$

Proof. With respect to [3, 4, 6] we have $f(\lambda A \times \mu B) = \sum_{\substack{r=0 \\ s=0}}^{d} \lambda^r \mu^s \chi_{rs}(A, B)$.

Since $f(tA) = tf(A)$ $(t \in \mathbb{Q}_+)$, we obtain $f(\lambda A \times \mu B) = \mu\chi_{01}(A, B) + \lambda\chi_{10}(A, B)$ and further $f(B) = \chi_{01}(A, B)$, $f(A) = \chi_{10}(A, B)$. $\lambda = \mu = 1$ presents $f(A \times B) = f(A) + f(B)$. $\square$

The question arises whether Theorem 1 holds in d dimensions. Following our argumentation in the same manner it would be interesting to know conditions where a motion-invariant valuation which is simply additive and additive in the sense of Minkowski vanishes on $P(E^d)$. The following theorem gives a possible condition.

THEOREM 3. Let χ be a simply additive valuation on $P(E^d)$, $d \geq 2$, which is additive in the sense of Minkowski, invariant under rigid motions and satisfying the inequality $\chi(A) \geq \alpha$ for all A contained in the unit cube W, where α is an existing real constant. In addition, there exists a function $K(E^d) \to \mathbb{R}$

which is continuous with respect to the Hausdorffmetric and $\chi + \psi\big|_{P(E^d)}$ is

monotonic. Then $\chi(A) = 0$ for all $A \in P(E^d)$.

We mention that we make no additional assumptions for ψ except the continuity and the fact that ψ is defined on $K(E^d)$.

Proof. If $\alpha \geq 0$ we get by usual argumentations $\chi(A)$ is monotonous on $P(E^d)$. Since $\chi(W) = 0$ for every cube, we obtain for $A \in P(E^d)$ and a cube W_0 with $A \subseteq W_0$: $0 \leq \chi(A) \leq \chi(W_0) = 0$ and $\chi(A) = 0$ follows.

If $\alpha < 0$ and $A \in P(E^d)$ with $0 \in A \subseteq W$ we get $-\chi(A) \leq -\alpha(-\alpha \geq 0)$ and since $\lambda A \subseteq W$ for $\lambda \in [0,1]$, we obtain setting $h(\lambda) = -\chi(\lambda A)$: $h(\lambda + \mu) = h(\lambda) + h(\mu)$ $(\lambda, \mu \in \mathbb{R}_+)$ and $h(\lambda) < M < \infty$ for all $\lambda \in [0,1]$ and so by a result of Ostrowski [5] $h(\lambda) = h(1)\lambda$, $\lambda \in \mathbb{R}_+$, which means $\chi(\lambda A) = \lambda\chi(A)$ for $A \in P(E^d)$ with $0 \in A \subseteq W$. Since $A \in P(E^d)$ is arbitrary, we have $A = (A \times \{-a\}) \times \{a\}$, $a \in A$, and $0 \in A \times \{-a\}$. As $\beta \in \mathbb{Q}_+$ such that $\beta(A \times \{-a\}) \subseteq W$, we obtain for $\lambda \in \mathbb{R}_+$, $\chi(\lambda\beta(A \times \{-a\})) = \lambda\chi(\beta(A \times \{-a\}))$ and it follows as χ is invariant under rigid motions: $\chi(\lambda A) = \lambda\chi(A)$. This equality also holds for $\lambda = 0$.

For $n \in \mathbb{N}$ and $\delta_1,\ldots,\delta_n$ rotations around the origin and $\lambda_i \geq 0$, $i = 1,\ldots,n$, $\sum_{i=1}^{n} \lambda_i = 1$ we define for $A \in P(E^d)$, $T(A) = \lambda_1 A^{\delta_1} \times \cdots \times \lambda_i A^{\delta_n}$. The transformation $T : P(E^d) \to P(E^d)$ is called "Drehmittelung." By Hadwiger [2] we know that for $A \in P(E^d)$ there exists a ball with center in the origin K and a sequence $(A_n)_{n \in \mathbb{N}}$ in $\mathcal{G} = \{T(A) \mid T"\text{Drehmittelung}"\}$ and $\lim_{n \to \infty} A_n = K$. By $B(a;r)$ we denote the ball with center a and radius r. Now let ε be an arbitrary positive real number. Because ψ is continuous, for $P \in P(E^d)$ arbitrarily chosen and with $r = \frac{1}{W_d} W_{d-1}(P)$, one can find real numbers r_0, r_1 with $0 < r_0 < r < r_1$ and $|\psi(B(0;r_i)) - \psi(B(0;r))| < \varepsilon$ $(i = 0,1)$.

Because of the theorem of Hadwiger mentioned above, the continuity of ψ and the inequality $r_0 < r < r_1$ we can find "Drehmittelungen" T_0,T,T_1 with the following properties:

$$T_0(W_0) \subseteq T(P) \subseteq T_1(W_1)$$

and

$$|\psi(T(P)) - \psi(B(0;r))| < \varepsilon$$

and

$$|\psi(T_i(W_i) - \psi(B(0;r_i))| < \varepsilon   (i = 0,1)$$

Since $X + \psi$ is monotonic we obtain

$$X(T_0(W_0)) + \psi(T_0(W_0)) \leq X(T(P)) + \psi(T(P)) \leq X(T_1(W_1)) + \psi(T_1(W_1))$$

Because X is invariant under rigid motions and additive in the sense of Minkowski and positively homogeneous for real numbers, we see that X is invariant under "Drehmittelungen." As X is simply additive $X(W_i) = 0$, $i = 0,1$, and so we get $\psi(T_0(W_0)) \leq X(P) + \psi(T(P)) \leq \psi(T_1(W_1))$ which shows

$$
\begin{aligned}
|X(P)| &\leq |\psi(T_1(W_1)) - \psi(T(P))| + |\psi(T_0(W_0)) - \psi(T(P))| \\
&\leq |\psi(T_1(W_1)) - \psi(B(0;r_1))| + |\psi(B(0;r_1)) - \psi(B(0;r))| \\
&\quad + |\psi(B(0;r)) - \psi(T(P))| + |\psi(T_0(W_0)) - \psi(B(0;r_0))| \\
&\quad + |\psi(B(0;r_0)) - \psi(B(0;r))| + |\psi(B(0;r)) - \psi(T(P))| \leq 6\varepsilon
\end{aligned}
$$

and so the theorem is proved. (The proof is a slight modification of a proof of Hadwiger in [2].) □

Another possibility to show that Theorem 1 holds in d dimensions ($d \geq 3$) is the solution of the following problem: Could one characterize the volume (up to a multiplicative constant) in $P(E^d)$ to be the only simple-additive, motion-invariant valuation f on $P(E^d)$ satisfying $f(A) \geq \alpha$ for all $A \subseteq W$ and a suitable $\alpha \leq 0$ (W unit cube)?

REFERENCES

1.   H. Hadwiger. Über beschränkte, additive Funktionale konvexer Polygone, *Publ. Math. Debrecen* 1(1949), 104-108.

2.   H. Hadwiger. *Vorlesungen über Inhalt, Oberfläche und Isoperimetrie,* Springer-Verlag, Berlin, 1957.

3.   P. McMullen. Valuations and Euler-type relations on certain classes of convex polytopes, *Proc. London Math. Soc.* 35(1977), 113-135.

4.   Ch. Meier. Multilinearität bei Polyederaddition, *Arch. Math.* 29(1977), 210-217.

5.   A. Ostrowski. Über die Funktionalgleichung der Exponentialfunktion und verwandte Funktionalgleichungen, *Ober. Dtsch. Math.-Ver.* 38(1929), 54-62.

6.   W. Spiegel. Ein Beitrag über additive, translationsinvariante, stetige Eikörperfunktionale, *Geom. Ded.* 7(1978), 9-19.

CONVEXITY THEOREMS FOR GENERALIZED
PLANAR CONFIGURATIONS

Jacob E. Goodman

Department of Mathematics
The City College
City University of New York
New York, New York

Richard Pollack

Department of Mathematics
Courant Institute of Mathematics
New York University
New York, New York

Arrangements of lines have been studied at least since Jakob Steiner [8] (see [6] for an excellent survey of the field up to 1971). Many of the problems that have been considered are really problems about the cell complexes determined by the arrangements. In an attempt to understand what is special, in regard to these problems, about the *straightness* of the lines, the notion of an arrangement of lines was generalized to the notion of an arrangement of pseudolines [7]: a pseudoline is a Jordan curve in RP^2 which does not separate the plane, and an arrangement of pseudolines is a finite collection of these with the property that any two meet exactly once (it then follows that where they meet they must cross), i.e., they meet each other pairwise the way straight lines do. Keeping in mind the nature of the problems that are being addressed, one calls two arrangements isomorphic if the corresponding cell complexes are isomorphic.

One criterion for evaluating a generalization such as this is whether the essential properties of the object being generalized (an arrangement of lines) remain true for the generalized object (an arrangement of pseudolines), and whether other properties—hopefully only inessential ones—are lost. The latter is true in this instance by virtue of the existence of *nonstretchable*

73

arrangements of nine or more pseudolines, i.e., arrangements not isomorphic
to any arrangements of lines [7]; on the other hand, it turns out that every
arrangement of eight or fewer pseudolines *is* stretchable [4].

Recently, in an effort to better understand the order and convexity properties of finite configurations of points, the notion of an "allowable sequence
of permutations" has been introduced [2]; this is a combinatorial object which
can be associated with a configuration of points in such a way that the order
and convexity relations which hold among the points of the configuration are
encoded in the allowable sequence associated with it. This is done as follows:
Let C be a configuration of points in E^2 labeled by the numbers $1,2,\ldots,n$ and
let L be a directed line which is not orthogonal to the direction determined by
any two points of C. Projecting the points of C orthogonally onto L determines
a permutation of the numbers $1,2,\ldots,n$. If L is now rotated counterclockwise,
the permutation changes as soon as L crosses one of the prohibited directions.

FIGURE 1 (a) & (b).

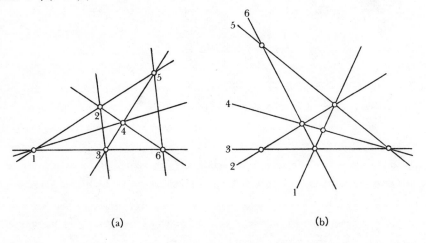

(a)   (b)

Letting L turn all the way around, we obtain a periodic sequence S of permutations associated with the configuration C; for example, the configuration
shown in Fig. 1(a) produces the sequence

$$\ldots 123456 \xrightarrow{23,56} 132465 \xrightarrow{246} 136425 \xrightarrow{136} 631425 \xrightarrow{14} 634125 \xrightarrow{125}$$

$$634521 \xrightarrow{345} 654321 \xrightarrow{65,32} 564231 \xrightarrow{642} 524631 \xrightarrow{631} 524136 \xrightarrow{41}$$

$$521436 \xrightarrow{521} 125436 \xrightarrow{543} 123456 \ldots$$

in which we have also indicated the move from each term to the next. This
sequence S satisfies the following conditions: (1) the move from any term
of S to the next is effected by reversing one or more disjoint substrings of
the former; (2) S is periodic; and (3) each pair i,j of indices is reversed
once and only once in every half period of S. Any sequence of permutations
with these properties is called an "allowable sequence."

Let S be an allowable sequence of permutations of $N = \{1,...,n\}$, and
let A and B be subsets of N. We say that A and B are *separated* in S if, in
some term of S, every element of A is (strictly) to the left of every element of B. (This is precisely what would happen if S arose from a configuration of points and the sets corresponding to A and B could be separated by a
line.) If $i \in N$ we say that i is in the *convex hull* of A if {i} and A are not
separated; this is equivalent to saying that in every term of S there is some
element of A which precedes (or equals) i. Finally, we consider two allowable
sequences to be equivalent if one can be obtained from the other by relabeling
the indices 1,2,...,n or by reversing the order in which the terms occur. (This
last operation corresponds either to having the line L rotate clockwise instead
of counterclockwise, or simply to reflecting the configuration in a line.)

Thus, an allowable sequence can be thought of as a generalized configuration of points. In fact, just as there are nonstretchable arrangements, it
turns out that there are also allowable sequences which do not arise from any
configurations of points [2]; such sequences are called nonrealizble.

These same sequences can also be used to describe arrangements of lines;
in fact they result in a somewhat finer classification than isomorphism. This
is accomplished as follows (for simplicity, we work in E^2 for the moment): Let
A be an arrangement of n lines labeled with the numbers 1,2,...,n and let L be
a directed "vertical" line to the "left" of all points of intersection of the
lines of A. The order in which the lines of A cross L determines a permutation
of {1,...,n}, and as L moves from left to right we obtain a half-period of an
allowable sequence. Since an allowable sequence is determined by any half-
period, we have associated such a sequence to the given arrangement of lines.
(For example, the arrangement shown in Fig. 1(b) also produces the sequence
above.) It turns out [3] that an allowable sequence is realizable by points
if and only if it is realizable by lines in the above sense; this fact extends
the usual duality principle in the real projective plane to include order and

convexity relations as well as incidence relations. For example, it turns out
that Helly's theorem for finite collections of polygons dualizes to

DUAL HELLY THEOREM. If $\Pi_1, \Pi_2, \ldots, \Pi_n$ are convex polygons in P^2 with a common
point such that for every i,j,k there is a line L_{ijk} which misses each of Π_i,
Π_j, and Π_k, then there is a line L which misses all the polygons $\Pi_1, \Pi_2, \ldots, \Pi_n$.

Although there are sequences which are not realizable by points and therefore not realizable by lines, every allowable sequence can nevertheless be realized by an arrangement of pseudolines; the sequence

$$\ldots 12345 \xrightarrow{12} 21345 \xrightarrow{34} 21435 \xrightarrow{35} 21453 \xrightarrow{14} 24153 \xrightarrow{24} 42153 \xrightarrow{15}$$

$$42513 \xrightarrow{13} 42531 \xrightarrow{25} 45231 \xrightarrow{45} 54231 \xrightarrow{23} 54321 \ldots$$

for example, is shown in [2] not to be realizable by points, but it is realizable by pseudolines in the sense depicted in Fig. 2. This observation, togethe

FIGURE 2.

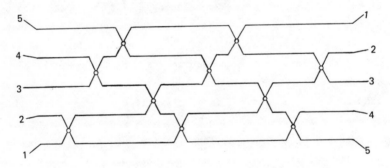

with a technical device known as the *Levi enlargement lemma*, permits us to
transfer many theorems about pseudoline arrangements to allowable sequences,
and vice versa. (The lemma says that a new pseudoline can be passed through
any two points which are not both on the same pseudoline of an arrangement,
so that a new arrangement results; see [6] for a proof.) The remainder of
this paper will be devoted to describing what the basic theorems about convex
sets look like when generalized to allowable sequences, and to arrangements of
pseudolines; their proofs will appear elsewhere [5].

The theorems that we shall consider are the following, in which A,B,...
are to be thought of as finite sets of points in the plane:

THEOREM A (Separation Theorem). If A and B cannot be separated by a line then there is a point common to their convex hulls.

THEOREM B (Radon's Theorem). Any set C containing at least four points can be partitioned into two sets A and B whose convex hulls meet.

THEOREM C (Helly's Theorem). If every three sets of a finite collection of sets $A_1, A_2, \ldots, A_n$ have a point common to their convex hulls, then there is a point common to the convex hulls of all of them.

THEOREM D (Carathéodory's Theorem). If x is in the convex hull of a set A with $|A| \geq 3$, then there is a subset B of A with $|B| = 3$ such that x is in the convex hull of B.

THEOREM E (Kirchberger's Theorem). If for every four points chosen from the union of two sets A, B the points of A can be separated from the points of B, then A can be separated from B.

In order for these to make sense for pseudoline arrangements, the statements "x is in the convex hull of A" and "A and B are separated" must first be interpreted for pseudolines; this is described below. A second problem which arises is that the first three of these theorems assert the existence of a new point, whereas our pseudoline arrangements (or allowable sequences, for that matter) do not live in any ambient space from which this new object can be chosen. We must therefore assert the existence of an *extension* of the pseudoline arrangement (or of the sequence) containing a new member with the desired properties. Hence for allowable sequences, thought of as generalized configurations of points, the theorems stated above should generalize to the following purely combinatorial assertions:

1. If S is an allowable sequence and A and B are sets of indices which are not separated then there is an extension S' of S and an index i such that in every term of S' some index of A and some index of B are each to the left of i.

2. If C is a subset of the set of indices of an allowable sequence S containing at least four elements, then there is an extension S' of S, a partition of C into two sets A and B, and an index i occurring in S' such that i is in the convex hulls of both A and B.

3. If S is an allowable sequence and $A_1, A_2, \ldots, A_n$ are sets of indices occur-
    ring in S such that for every i,j,k there is an index x_{ijk} which belongs
    to the convex hull of each of A_i, A_j, and A_k, then there is an allowable
    sequence S' which extends S and contains an index x which belongs to the
    convex hull of each A_i,   i = 1,2,...,n.

4. If S is an allowable sequence and A is a set of at least three indices
    occurring in S such that some index x is in the convex hull of A, then
    there are indices i,j,k in A such that x is in the convex hull of {i,j,k}.

5. If S is an allowable sequence and A,B are sets of indices occurring in S
    such that for every $A' \subset A$, $B' \subset B$ with $|A' \cup B'| \leq 4$ there is a term of
    S in which the indices of A' are to the left of those of B', then there
    is a term of S in which every index of A is to the left of every index of
    B.

Notice that for points $P, P_1, P_2, \ldots, P_n$ in the euclidean plane, the fact that
P is in the convex hull of $P_1, \ldots, P_n$ can be described by saying that a line L,
which starts out with all the points $P, P_1, \ldots, P_n$ on one side of it, cannot be
moved continuously to meet P without also meeting one of $P_1, \ldots, P_n$. Since E^2
is just P^2 minus a line, for points $P, P_1, \ldots, P_n$ in P^2 none of which lie on a
distinguished line L, we can say that P is in the L-*convex hull* of $P_1, \ldots, P_n$
if L cannot be moved continuously to meet P without also meeting one of $P_1, \ldots,$
P_n. Now that we are in P^2, the duality principle suggests that we define a
line L to be in the P-*convex hull* of lines $L_1, \ldots, L_n$ (assuming that none of
these lines contains P) if P cannot be moved continuously to meet L without
also meeting one of $L_1, \ldots, L_n$. Notice that this is the same as saying that
the line L does not meet the (open) P-cell of the arrangement $\{L_1, \ldots, L_n\}$,
i.e., the P-component of the complement of $L_1 \cup \cdots \cup L_n$. It is now clear
how to define the statement "L is in the P-convex hull of $L_1, \ldots, L_n$," where
$L, L_1, \ldots, L_n$ are pseudolines belonging to an arrangement. In a similar way we
can dualize the idea of separating two sets of points by a line in E^2, to obtain the following definition: If B and C are subsets of an arrangement A of
pseudolines and P is a point on none of the pseudolines of A, then the sets B
and C are P-*separated* if there is a point Q and a pseudoline L through P and
Q which extends the arrangement A (such pseudolines exist by the Levi enlargement lemma) such that L meets the pseudolines of B in one component of L \ {P,Q}
and meets those of C in the other component.

It is now easy to see that the five theorems above should be generalized
as follows for pseudoline arrangements:

1'. If B and C are subsets of an arrangement A of pseudolines, and P is a
     point which meets none of the pseudolines of A, and if B and C are not
     P-separated, then there exists an extension A' of A and a pseudoline

$L \in A'$ which avoids both the P-cell of the arrangement B and the P-cell of the arrangement C.

2'. If B is a set of at least four pseudolines chosen from an arrangement A and P is a point which meets no pseudoline of A, then there is a partition of B into $B_1 \cup B_2$ and a pseudoline L which extends the arrangement A such that L avoids the P-cell of B_1 as well as the P-cell of B_2.

3'. If $A_1, \ldots, A_n$ are subsets of an arrangement A none of whose pseudolines passes through a point P, such that for each i,j,k there is a pseudoline $L_{ijk} \in A$ which avoids the P-cell of each of A_i, A_j, A_k, then A can be extended to an arrangement A' containing a pseudoline L which avoids the P-cell of each A_i $(i = 1, \ldots, n)$. (Note the resemblance to the "Dual Helly Theorem" mentioned above.)

4'. If B is a subset of an arrangement A none of whose pseudolines meets P, and $L \in A$ avoids the P-cell of B, then there are pseudolines $L_1, L_2, L_3 \in B$ such that L avoids the P-cell of $\{L_1, L_2, L_3\}$.

5'. If A and B are subsets of an arrangement none of whose pseudolines meets P, and for every choice of ≤ 4 pseudolines from $A \cup B$ those of A are P-separated from those of B, then A and B are P-separated.

Although we do not prove these theorems here, a few features of the proofs are worth noting. Theorem 2' is essentially a consequence of the Levi enlargement lemma, modulo the exercise of some caution about extending arrangements. 3' follows from 2' in the usual way that Helly follows from Radon. Theorem 4' is proven by a purely combinatorial argument, which applies equally well to Theorem 4. All the unprimed theorems are consequences of the primed theorems, since every allowable sequence has a realization by an arrangement of pseudolines. (On the other hand, an independent combinatorial proof of Theorems 2 and 3 is also possible, using a combinatorial version of the Levi enlargement lemma.) The surprise is Theorem 5' (which easily implies 1'): although it appears that the theorem is of an essentially combinatorial nature, the only proof we have found so far uses a fair amount of topological machinery, in particular Helly's topological theorem in three dimensions. It would be nice to have a purely combinatorial proof of this theorem, i.e., a direct proof of Theorem 5.

Because of the connection established in [1] between arrangements of pseudohyperplanes and oriented matroids, the theorems stated here are also theorems about rank three oriented matroids. Extending these results to matroids of higher rank would be identical with extending the theorems above to higher dimensions, which would obviously be desirable in its own right. A key step in this direction would be to prove an analogue of the Levi enlargement lemma for arrangements of pseudohyperplanes.

REFERENCES

1.   J. Folkman and J. Lawrence. Oriented matroids, *J. Comb. Theory*, Series B 25(1978), 199-236.

2.   J. E. Goodman and R. Pollack. On the combinatorial classification of nondegenerate configurations in the plane, *J. Comb. Theory*, Series A 29(1980), 220-235.

3.   J. E. Goodman and R. Pollack. A theorem of ordered duality, to be published in *Geometriae Dedicata*.

4.   J. E. Goodman and R. Pollack. Proof of Grünbaum's conjecture on the stretchability of certain arrangements of pseudolines, *J. Comb. Theory*, Series A 29(1980), 385-390.

5.   J. E. Goodman and R. Pollack. Helly-type theorems for pseudoline arrangements in P^2, to be published in *J. Comb. Theory*, Series A.

6.   B. Grünbaum. *Arrangements and Spreads*, Amer. Math. Soc., Providence, R. I., 1972.

7.   F. Levi. Die Teilung der projektiven Ebene durch Gerade oder Pseudogerade, *Ber. Verh. sächs. Ges. Wiss. Leipzig Math. Phys. Kl.* 78(1926), 256-267.

8.   J. Steiner. Einige Gesetze über die Teilung der Ebene und des Raumes, *J. Reine Angew. Math.* 1(1826), 349-364.

REMARKS ADDED IN PROOF

Since this paper was written, the authors have discovered a counterexample to the "Levi enlargement lemma" in dimension ≥ 3 [Three points do not determine a (pseudo-)plane, to be published in *J. Comb. Theory*, Series A], which shows that our hoped-for method of extending the theorems above to higher dimensions is doomed to failure. Moreover, A. Mandel [private communication] has an example which shows that the generalization of the separation theorem (1') to higher dimensions is false, for a reason related to the failure of the Levi enlargement lemma.

On the other hand, R. Cordovil [Sur un théorème de séparation des matroïdes orientés de rang trois, preprint] has successfully used the techniques of oriented matroids to give an alternate proof of the Kirchberger theorem (5') for pseudoline arrangements, and to generalize the Carathéodory theorem (4') to higher dimensions. Since his proof of Kirchberger makes use of the Folkman-Lawrence theorem on the representability of an oriented matroid by an arrangement of pseudohemispheres, a result which relies heavily on topological arguments of its own, it is not really a "purely combinatorial" proof of the sort we envisioned above, i.e., a direct proof of (5); his results do suggest, however, that suitably stated higher dimensional generalizations of (2'), (3'), and (5') may yet be obtainable.

IS THERE A KRASNOSELSKII THEOREM FOR FINITELY STARLIKE SETS?

Bruce B. Peterson

Department of Mathematics
Middlebury College
Middlebury, Vermont

If S is a subset of E^d and x and y are points of S, then x *can see* y via S if the segment $xy \subseteq S$. The set S is *starlike* if there is a point x in S which can see every point of S via S. The set S is *finitely starlike* if for every finite subset T of S, there is a point t of S such that t can see every point of T via S.

Krasnoselskii [2] proved that for compact sets, if every set of at most n + 1 points can see a common point, then S is starlike. Of course, then any compact finitely starlike set is starlike.

Nonstarlike finitely starlike sets abound:

EXAMPLE 1. In E^2 the region bounded by the graph of $y = \dfrac{1}{1 - x^2}$ and its asymptotes,

$$S_1 = \{(x,y) \mid 0 < x < 1, y \leq \frac{1}{1 - x^2} \} \cup \{(x,y) \mid x = \pm 1\}$$

is finitely starlike but not starlike.

FIGURE 1.

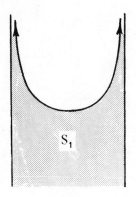

EXAMPLE 2. In E^2, the set $S_2 = \{(x,y) \mid y \leq \sin x\}$ is finitely starlike but not starlike.

EXAMPLE 3. If S is any convex body in E^d, $d \geq 3$, and T is any finite subset of S, then S − T is finitely starlike. In general, if $\{F_i\}$ is a countable collection of flats of dimension $\leq d - 2$, then S− $\cup$ (S $\cap$ F_i) is finitely starlike.

EXAMPLE 4. The closed set S_3 (Fig. 2, distorted) is starlike. $S_3 - \{0\}$, however, is not starlike but is finitely starlike. The set is formed by choosing a pencil $\{m_n\}$ of lines on 0 (y = x/n will do) which converges to

FIGURE 2.

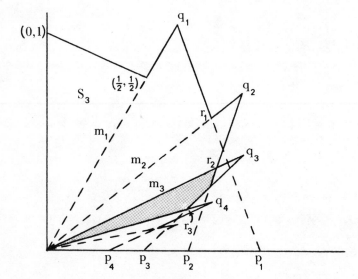

the positive x-axis, and a decreasing sequence of points p_n on the positive x-axis converging to 0. We form a polygon as follows: Begin with
vertices $(0,0)$, $(0,1)$, $(\frac{1}{2},\frac{1}{2})$. Then on $y = x/2$, choose a vertex beyond $(\frac{1}{2},\frac{1}{2})$,
call it q_1. Let $r_1 = q_1 p_1 \cap m_2$. On m_2 choose a vertex beyond r_1; call it
q_2. On Let $r_2 = q_2 p_2 \cap m_3$, and choose q_3 on m_3 beyond r_2, etc. Any finite
collection of points in S_3 can see an entire quadrilateral (the common visibility set for $\{q_1,q_2,q_3\}$ is shaded in the figure), but the entire set is
starlike from zero alone.

We would like to find some sort of Krasnoselskii theorem for finitely
starlike sets. The set S_4^2 consisting of a closed square with an open diagonal removed is not finitely starlike since the four vertices cannot see a
common point, but every three points of S_4^2 can see a common point.

FIGURE 3.

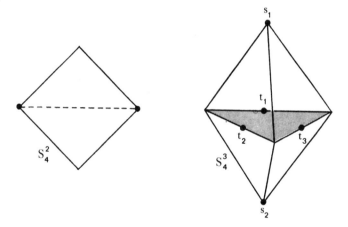

In E^d, the suspension of a $(d - 1)$-simplex with the relative interior of the
simplex deleted

$$S_4^d = (\Sigma \sigma^{d-1}) - \text{rel int}(\sigma^{d-1})$$

fails to be finitely starlike because the set T consisting of the barycenters
of the $(d - 2)$-faces of σ^{d-1} plus the two suspending vertices is a set of $d +$
2 members for which no commonly visible point exists. Yet for any collection
of $d + 1$ points in S, there is a point from which all the points can be seen.
The set S_4^3 is pictured for E^3.

Proofs of Krasnoselskii's Theorem apply Helly's Theorem to show that, if
V_x is the set of all points which x can see via S, then $\bigcap_{x \in S} \text{conv } V_x \neq \emptyset$,

and that, in fact, $\bigcap\limits_{x \in S} \text{conv } V_x = \bigcap\limits_{x \in S} V_x$. For finitely starlike sets, even closed finitely starlike sets, this approach is not readily generalizable. Indeed in S_1 the convex hulls of visibility sets may intersect in points outside of S_1, so that passing from the intersection of three visibility sets to the intersection of four cannot be accomplished via Helly's Theorem.

FIGURE 4.

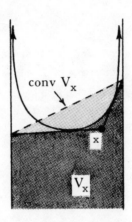

REFERENCES

1. L. Danzer, B. Grunbaum, and V. Klee. Helly's Theorem and its relatives, *American Mathematical Society Symposium on Convexity, Seattle, Proceedings of the Symposium in Pure Mathematics*, Volume 7, 1963.

2. M. A. Krasnoselskii. *Sur un critère pour qu'un domain soit etoilé, Mat. Sb. (N.S.)* 19(61)(1946), 309-310.

3. F. A. Valentine. *Convex Sets*, McGraw-Hill, New York, 1964.

CONVEX CAUSTICS FOR BILLIARDS IN R^2 AND R^3

Philip H. Turner*

Department of Mathematics
Louisiana State University
Baton Rouge, Louisiana

1. INTRODUCTION

We are going to play idealized billiards inside some compact convex bodies of R^2 and R^3. We will represent the billiard ball as a point which moves at unit speed in a straight line except when it hits the boundary of the containing body. At the boundary the ball rebounds elastically according to the law of equal angles of incidence and reflection.

This idealized model for billiards inside a compact convex body forms the intuitive basis for the definition of a dynamical system which has recently received considerable mathematical attention. (We invite the reader to peruse the appended Bibliography on Billiards. The author would appreciate receiving notice of any possible additions to this bibliography.) This literature on billiards may be regarded as a memorial to the great American mathematician G. D. Birkhoff who seems to have been the first to popularize this dynamical system through his books and articles. Let us also note here that this dynamical system has an alternative intuitive description in terms of a focused light ray reflecting within the silvered boundary of a body. When this interpretation of the dynamics is used, our topic of study is included within the subject known as geometric optics.

Current affiliation: Department of Systems, United Illuminating Company, New Haven, Connecticut.

A formal description of this dynamical system in terms of a group of measure-preserving transformations on phase space seems inappropriate for this paper. The interested reader may consult Turner [10, pp. 213-221] for such a formal description. It will suffice for us to play an intuitive game of billiards. Let us note, furthermore, that we will not require a special reflection convention at corners, i.e., boundary points through which there passes more than one supporting hyperplane of the given convex body. We will play billiards inside smooth convex bodies, i.e., convex sets with interior whose boundaries have no corners.

In this paper we will be engaged in a search: a search for objects we call convex caustics for billiards. We will begin our search in the plane R^2, where in one sense the search has been completed through the author's efforts. We will then search for caustics in the space R^3, where the search has barely begun.

2. CONVEX CAUSTICS FOR BILLIARDS IN R^2

Let us start with a description of the objects of our search.

DEFINITION 1. Suppose Γ is a smooth compact convex body in R^2. A proper closed convex subset K of Γ will be called a *convex caustic for billiards inside* Γ if any billiards trajectory which travels in either direction along a supporting line of K before relection at bd Γ remains on a supporting line of K after reflection.

REMARK: This definition both generalizes and restricts a similar notion of a caustic for billiards found in Sinai [6, p. 87]. It restricts attention to caustics which are convex subsets of Γ so that we may replace tangents in Sinai's definition by supporting lines. This replacement enables us to consider convex sets which do not have differentiable boundary curves.

This notion of a caustic for billiards should not be confused with certain plane curves which are also (unfortunately) known as caustic curves. These latter caustic curves are constructed from a given curve S and a fixed point F called the radiant point. One considers the family of rays (or, equivalently, billiards trajectories) passing through F and reflected by the curve S. Then the caustic curve of S with F as radiant point is the envelope of the family of these once-reflected rays (cf. Lockwood [4, p. 183]). Such

a curve is not a caustic for billiards inside S, however, since the reflective property of a caustic for billiards must hold for rays along *any* supporting line, not just those which pass through the given point F.

EXAMPLES: The classical example of a convex caustic for billiards is apparently due to Jacobi, although we do not have a specific reference to Jacobi's
work. If Γ is the convex hull of an ellipse in R^2 focused at F_1 and F_2,
then the convex hull of any confocal ellipse properly contained in Γ is a
convex caustic for billiards inside Γ. (See Fig. 1.) We refer the interested
reader to Sinai [6, p. 86] for a proof of this classical result.

It is not hard to find some elementary examples of convex caustics for
billiards. If K is a singleton point P, for example, then choose Γ to be
any disk of radius $r > 0$ centered at P. Any supporting line of K is just a
line through P. A billiards trajectory travelling along a supporting line,
then, travels along a diameter of Γ and must return along the same diameter
after reflection at the boundary. Hence K is a convex caustic for billiards
inside Γ.

If K is a line segment F_1F_2, then choose Γ to be the convex hull of any
nondegenerate ellipse focused at F_1, F_2. Any supporting line of K is just a
line through F_1 or F_2 and the focal reflecting property of an ellipse guar-
antees that a billiards trajectory passing through one focus passes through
the other focus after reflection. Thus any line segment is a convex caustic
for billiards inside such a Γ.

Suppose now that K is a simplex with vertices at F_1, F_2, F_3. If we wrap
a loop of inelastic string around K whose length is greater than the perimeter
of the simplex K, then by using a pencil to pull the string taut around K we

FIGURE 1

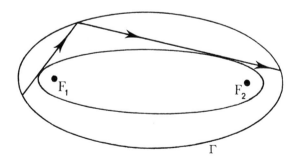

Γ

FIGURE 2

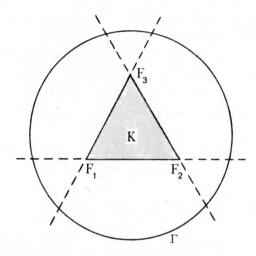

may trace a simple closed curve γ enclosing K. (See Fig. 2.) This curve γ
may be seen to be a smooth union of 6 elliptical arcs focused at combinations
of the three vertices F_1, F_2, F_3 taken two at a time. Any billiard trajectory
which travels along a supporting line of K passes through at least one vertex of K. Because of the focal reflective property of ellipses, such a trajectory must pass through the other focus of the reflective elliptical arc.
Since this other focus is also a vertex of K, it is clear that the trajectory
remains on a supporting line of K after its reflection. Hence K is a convex caustic for billiards in the bounded component of γ.

    A similar argument shows that if K is any convex polygon in R^2 with n
extreme points, where n is an integer greater than 3, then the string curve
γ is once again a smooth union of m elliptical arcs where $n \leqslant m \leqslant 2n$ and K
is a convex caustic for billiards in the bounded component of γ.

    Now if K is any compact convex body in R^2 we can approximate K by convex
polygons. It is not too surprising then that the string curve γ, constructed
as above for the simplex, will provide the proper reflective container for K
so that K is a convex caustic for billiards inside this container. This result is one half of our Theorem 2 below. The elementary examples above moti-
vate the following definitions.

STANDING HYPOTHESIS. Let K be a compact convex set in R^2.

DEFINITION 2. Let us denote the 1-dimensional surface area of bd K by $SA_1(K)$. The R^2-*embedded profile* of K determined by the constant λ, where $\lambda > SA_1(K)$, is $\{P \in R^2 \mid SA_1(\text{conv}(K \cup P)) = \lambda\}$. We will denote this profile as γ_λ.

REMARK. If $\text{conv}(K \cup P)$ denotes the convex hull of $K \cup P$, such convex sets being called *cap bodies (Kappenkörper)* in the older literature (cf. Bonnesen-Fenchel [1, pp. 17-19]), then $SA_1(\text{conv}(K \cup P))$ may be interpreted as the length of the inelastic string wrapped around K and pulled taut at P (Fig. 3).

FIGURE 3

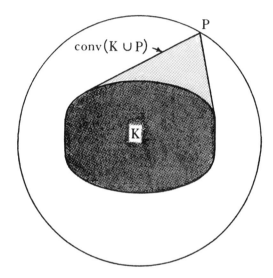

DEFINITION 3. We call the function $f(P) = SA_1(\text{conv}(K \cup P))$ the *string length function* of K at P.

REMARK. It is clear that the profile γ_λ is simply the level curve of the string length function. It can be shown that this function is a convex function and, consequently, that the sublevel set $\Gamma_\lambda = \{P \in R^2 \mid SA_1(\text{conv}(K \cup P)) \leq \lambda\}$ must be convex. This functional approach to studying profiles is found in Stoll [7, pp. 48-50, 54]. When the present author began studying profiles, he was unaware of Stoll's work and developed an independent appraoch. This more geometric approach is outlined below, partly because it has the advantage of showing Γ_λ is smooth and rotund as well as convex, properties of Γ_λ which Stoll does not appear to have considered.

MONOTONICITY LEMMA. Suppose H is any supporting line to K. As one travels
along H in a direction away from K ∩ H, the string length function is strictly
increasing and continuous.

Proof. K ∩ H is either a closed line segment AB or a point A = B. If
P ≠ Q are points of H \ AB such that Q is between P and A but B is not, then
we have conv(K ∪ Q) $\subsetneq$ conv(K ∪ P). (See Fig. 4.) The strict monotonicity of
the surface area functional SA_1 on compact sets in R^2 (cf. Bonnesen-Fenchel
[1, p. 47]) implies that $f(Q) = SA_1(conv(K ∪ Q)) < SA_1(conv(K ∪ P)) = f(P)$.
This shows that f is strictly increasing as we travel along H away from AB.
The continuity of f follows from the continuity of SA_1 with respect to the
Hausdorff metric applied to the collection of cap bodies, conv(K ∪ P). □

FIGURE 4

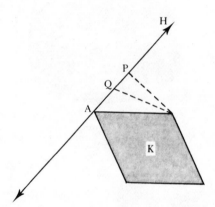

EXISTENCE LEMMA. Suppose H is any supporting line of K. For any choice of
$\lambda > SA_1(K)$, there exist exactly two points of γ_λ which lies on H.

Proof. Use the Monotonicity Lemma above and the Intermediate Value Theorem. □

REMARK. Suppose P ∉ K and that conv(K ∪ P) has nonempty interior. Then the
set bd conv(K ∪ P) consists of three parts. Two of these are line segments
along the two supporting lines to K passing through P and the third is the arc
shared by bd K and bd conv(K ∪ P). (See Fig. 5.) Since bd K ∩ bd conv (K ∪ P)
is compact, this shared arc is closed with two distinct endpoints. Since
conv(K ∪ P) has nonempty interior, its boundary is a simple closed curve which
may be oriented in the usual counterclockwise sense. Let A denote the end-

point of the shared arc which is prior from P in this sense and let Z denote
the other (later) endpoint. Then A and Z are the first and last points, respectively, of bd K which touch the string of length λ wrapped around K and
pulled taut at P. This intuitive picture motivates the following definition.

DEFINITION 4. Let $P \notin K$. If conv($K \cup P$) has nonempty interior, orient the
set bd conv($K \cup P$) in the counterclockwise sense and let A and Z denote the
endpoints of bd conv($K \cup P$) $\cap$ bd K so that A is prior from P in this sense.
Then we will call A the *lead contact point* of P and Z the *trail contact point*
of P. If conv($K \cup P$) has empty interior, let A = Z be the point of K closest
to P, and let A = Z be called both the lead and trail contact point of P.
Furthermore, we will call the angle APZ, in either case, the *string angle* at P.

FIGURE 5

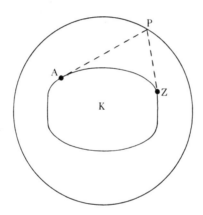

REMARK. This rather specialized terminology facilitates the formulation of
the next lemma which is the main tool in our proof of Theorem 1.

COMPARISON LEMMA. Let $P \in \gamma_\lambda$ with lead contact point A and trail contact
point Z. Let B be the bisector of the string angle APZ and let T be the normal line to B passing through P. Let T^+ denote the half plane determined by
T containing K. Then we can find two ellipses E_I and E_O, both contained in
T^+ and tangent to T at P such that

    (i)   for all $M \in E_I$, $SA_1(\text{conv}(K \cup M)) \leqslant \lambda$

    (ii) for all $L \in E_O$, $SA_1(\text{conv}(K \cup L)) \geqslant \lambda$

Proof. Case 1: Assume that conv(K ∪ P) has nonempty interior. Project K orthogonally onto B and let XY denote the closed line segment on B
which is the image of K under this projection. Let Y denote the closer of the
endpoints to P. Construct E_I as follows: Let W be the point on segment PY
which lies 1/3 of the way to Y from P. Construct line NW parallel to line T
through W and let F_1 = line NW ∩ line PA and F_2 = line NW ∩ line PZ. Let E_I
be the ellipse focused at F_1 and F_2 passing through P, as in Fig. 6. It is
clear that E_I ∩ K = ϕ. We note that T is tangent to E_I at P, since we have
focused E_I along lines PA and PZ respectively.

FIGURE 6

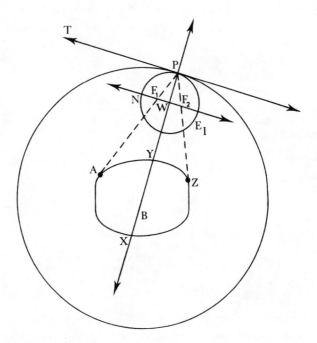

We construct E_0 as follows: Let V be the point on ray PX which lies
$\lambda/2$ units further from P than does X. Construct line NV parallel to line T
through V and let F_3 = line NV ∩ line PA and F_4 = line NV ∩ line PZ. Let E_0
be the ellipse focused at F_3, F_4 passing through P. See Fig. 7.

We omit the proofs of the inequalities (i) and (ii) here: the interested
reader may consult the author's doctoral thesis [10, pp. 149-161] for further
details.

Case 2: Assume that conv(K ∪ P) has empty interior. Then conv(K ∪ P)
is a closed line segment and A = Z is the point of K closest to P. Let V be

FIGURE 7

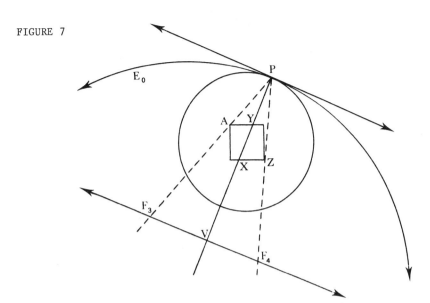

the point of K which is furthest from P. Let E_I be the circle centered at
A and passing through P and let E_0 be the circle centered at V and passing
through P. (See Fig. 8.) Clearly E_I and E_0 satisfy the desired properties. $\square$

REMARK. It should be apparent that E_I and E_0 may be chosen in a somewhat
arbitrary manner. The question naturally arises of how large (small) E_I (E_0)
can be and still satisfy the desired inequality.

FIGURE 8

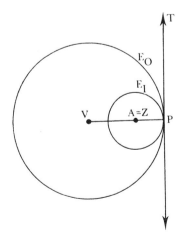

We are now ready to state and prove Theorem 1 which we first announced in Turner [8].

THEOREM 1. Any R^2-embedded profile γ_λ of K is the boundary of a compact convex body $\Gamma_\lambda \supseteq K$. Γ_λ is both smooth and rotund. If $P \in \gamma_\lambda$ and B is the bisector of the string angle APZ and T is the normal line to B through P (as in the Comparison Lemma above), then T is the unique supporting line to Γ_λ at P.

Proof. Given any $P \in \gamma_\lambda$, we can construct T. Index T by P and let T_P^+ denote the half-plane determined by T_P which contains K. Let $\Gamma_\lambda = \cap \{T_P^+ \mid P \in \gamma_\lambda\}$. Then clearly Γ_λ is convex and closed. $\Gamma_\lambda \supseteq K$, since $T_P^+ \supseteq K$ for all $P \in \gamma_\lambda$. Γ_λ is bounded, since $\Gamma_\lambda \subseteq B(\lambda, K)$, so Γ_λ is compact. By exploiting the Comparison Lemma, it may be shown that $\Gamma_\lambda = \{Q \in R^2 \mid f(Q) \leqslant \lambda\}$. Thus γ_λ is bd Γ_λ and the first statement is proved.

We can see that Γ_λ is smooth and rotund by using the Comparison Lemma. If $\Lambda \neq T$ is a line through P, then $\Lambda \cap$ int conv $E_I \neq \phi$. But int conv $E_I \subseteq$ int Γ_λ, so Λ cuts Γ_λ. Hence only T is a supporting line of Γ_λ at P. Furthermore, if $P^1 \neq P$ is some point on T, P^1 lies in the unbounded component of E_0. Hence conv$(K \cup P^1) > \lambda$ and $P^1 \notin \Gamma_\lambda$. Thus Γ_λ must be rotund as well as smooth. □

REMARK. Since $\Gamma_\lambda = $ conv γ_λ is a smooth compact convex body in R^2 by Theorem 1, we are now prepared to search for a convex caustic for billiards inside Γ_λ. Of course, K is this caustic as one half of Theorem 2 shows. However, the second half of Theorem 2 shows that possession of a convex caustic for billiards is sufficient to characterize conv γ_λ, a result we announced in Turner [9]. This result depends on a proposition found in Stoll [7, p. 60], which we state below.

PROPOSITION (Stoll). Suppose $f(P) = SA_1(\text{conv}(K \cup P))$ where $P \notin K$. Then f: $R^2 \setminus K \to R$ is C^1 with

$$\text{grad } f(P) = \frac{P - A}{\|P - A\|} + \frac{P - Z}{\|P - Z\|}$$

where A and Z are the lead and trail contact points of P, respectively.

Proof. We refer to Stoll [7, p. 60], or to Turner [10, pp. 185-194] for a more detailed proof.

REMARK. Since $\frac{P - A}{\|P - A\|}$ and $\frac{P - Z}{\|P - Z\|}$ are unit vectors, their sum grad f(P) lies on the bisector B of the string angle APZ. See Fig. 9. By Theorem 1, Γ_λ is smooth, so when its boundary γ_λ is parametrized by arclength s, $\frac{d}{ds} \gamma_\lambda(P)$ exists, and $\frac{d}{ds} \gamma_\lambda(P)$ is a unit tangent vector lying along T. Since B and T are normal lines, $\frac{d}{ds} \gamma_\lambda(P)$ and grad f(P) are orthogonal, as they should be. For the unit tangent vector of a level curve of a C^1 function f must be orthogonal to grad f.

THEOREM 2. Suppose K is any compact convex set in R^2 contained in the interior of a smooth compact convex body $\Gamma \subseteq R^2$. Then K is a convex caustic for billiards inside Γ if and only if bd Γ is an R^2-embedded profile of K.

Proof. ($\Leftarrow$) Suppose bd Γ is the R^2-embedded profile of K determined by $\lambda > SA_1(K)$. Then $\Gamma = \Gamma_\lambda$. Suppose a billiards trajectory travels along a

FIGURE 9

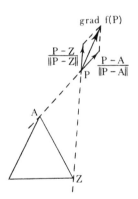

supporting line to K before reflection at $P \in$ bd $\Gamma_\lambda = \gamma_\lambda$. The unique supporting line T to Γ_λ at P is the normal to the bisector B of string angle APZ by Theorem 1. Let C be a point of $B \cap T^+ \setminus P$. The angle of incidence is thus a complementary angle of angle APC, as in Fig. 10. After the reflection at P, the angle of reflection must also be a complementary angle of angle APC. But

FIGURE 10

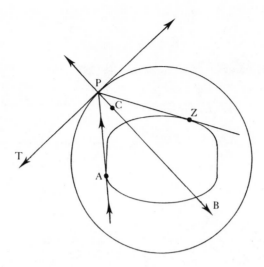

since B bisects angle APZ, the measure of angle APC is equal to the measure
of angle ZPC. Hence the angle of reflection must be a complement of angle
ZPC. This forces the billiards trajectory to travel along ray PZ after reflection at P. Since ray PZ ⊆ supporting line PZ of K, we have shown that K
is a convex caustic for billiards in Γ_λ.

   (⇒) Suppose K is a convex caustic for billiards in Γ. Since Γ is a
compact convex body in R^2, bd Γ may be represented as the image of a rectifiable simple closed curve which may be parametrized by arclength s. Denote
this boundary curve as γ. Then $\gamma\colon [0, SA_1(\Gamma)] \to$ bd Γ is a C^1 function of its
parameter s, since Γ is smooth and $\frac{d\gamma}{ds}$ is a unit tangent vector which lies a-
long the unique supporting line T to Γ at $\gamma(s)$. Consider the composed function

$$f(\gamma(s)) = SA_1(\text{conv}(K \cup \gamma(s)))$$

Since f is also a C^1 function by Stoll's proposition, $\frac{d}{ds} f(\gamma(s))$ exists and
equals grad $f(\gamma(s)) \cdot \frac{d}{ds} \gamma(s)$ by the Chain Rule. We claim that for all $s \in$
$[0, SA_1(\Gamma)]$, grad $f(\gamma(s)) \cdot \frac{d}{ds} \gamma(s) = 0$.

   K is a convex caustic for billiards in Γ by hypothesis. Given $\gamma(s)$,
draw the supporting lines to K and let A(s) and Z(s) denote the lead and
trail contact points of $\gamma(s)$, respectively. The law "angle of incidence e-

quals angle of reflection" at $\gamma(s)$ implies that

$$\frac{d\gamma}{ds} \cdot \frac{\gamma(s) - A(s)}{\|\gamma(s) - A(s)\|} = - \frac{d\gamma}{ds} \cdot \frac{\gamma(s) - Z(s)}{\|\gamma(s) - Z(s)\|}$$

(See Fig. 11.)   Hence

$$\frac{d\gamma}{ds} \cdot \left[\frac{\gamma(s) - A(s)}{\|\gamma(s) - A(s)\|} + \frac{\gamma(s) - Z(s)}{\|\gamma(s) - Z(s)\|} \right] = 0$$

By Stoll's proposition,

$$\text{grad } f(\gamma(s)) = \frac{\gamma(s) - A(s)}{\|\gamma(s) - A(s)\|} + \frac{\gamma(s) - Z(s)}{\|\gamma(s) - Z(s)\|}$$

Hence $\frac{d}{ds} f(\gamma(s)) = 0$ as claimed and $f(\gamma(s)) \equiv \lambda$ (constant). So $\gamma(s) \in \gamma_\lambda$ for all $s \in [0, SA_1(\Gamma)]$ and bd $\Gamma \subseteq \gamma_\lambda$. It is clear that $\gamma_\lambda \subseteq$ bd Γ, for otherwise bd Γ cannot be a simple closed curve. Thus $\gamma_\lambda = $ bd Γ. $\square$

FIGURE 11

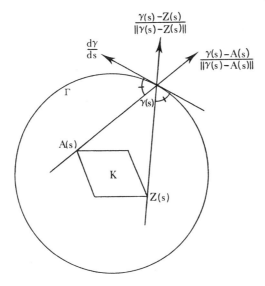

HISTORICAL REMARK. Theorem 2 is a generalization of a statement (made without proof) concerning convex caustics for billiards with differentiable boundary curves found in Sinai [6, p. 87]. The author found this statement interesting, for it asserted that a certain metric invariant (which we have interpreted as SA_1 or conv($K \cup P$) here) was related to a reflective property; namely, possession of a convex caustic. The author is indebted to his advisor, Professor Walter H. Gottschalk, for pointing out that the metric invariant could be interpreted as the length of a closed loop of inelastic string wrapped around the caustic. This interpretation led to the conclusion that the smoothness assumption on the boundary of the caustic was probably unnecessary, and it suggested, of course, the close connection between profiles and ellipses which the reader has seen us exploit in several ways.

AN OPEN PROBLEM. In one sense, Theorem 2 provides a complete answer to the search for convex caustics in the plane: given any compact convex set K, K is a convex caustic for billiards inside any member of its one parameter family of profiles $(\gamma_\lambda \mid \lambda > SA_1(K))$. In another sense, however, Theorem 2 is not very satisfactory. If we are given any smooth compact convex body Γ, how can we tell whether Γ possesses a convex caustic K for billiards inside Γ? By Theorem 2, we know that bd Γ must be an R^2-embedded profile of K, if K exists, but we do not know how to distinguish profiles intrinsically from boundaries of general smooth compact convex bodies in R^2. Theorem 1 provides a necessary condition; namely, Γ must be rotund. But it seems highly unlikely to this author that this condition should also be sufficient. Unfortunately, a "counter-example" to demonstrate this nonsufficiency fell through as this paper was being readied, so it is still possible that rotundity is sufficient. The problem, then, is to find some converse to Theorem 1.

This problem has been partially addressed by the Russian mathematician V. F. Lazutkin in [3], where he states the following theorem.

THEOREM (Lazutkin). Suppose that Γ is a smooth compact convex body in R^2. Let $\rho(s)$ be the radius of curvature function of a boundary curve of Γ parametrized by arclength s. If $0 < m < \rho(s) < M < \infty$ for all $s \in [0, SA_1(\Gamma)]$ and if $\rho(s)$ is C^{553}, then there exists a discontinuous family of convex caustics for billiards inside Γ with boundaries contained in a small neighborhood of bd Γ.

Proof. We refer the interested reader to Lazutkin's paper [3].

REMARK. It is easy to see that the profiles of a simplex have a discontinuity in their curvature function wherever they cross an extended side of the simplex. Thus there are elementary examples of profiles where $\rho(s)$ is not even continuous! This shows that Lazutkin's C^{553} condition is very far from necessary. We note, however, that $0 < \rho(s) < M < \infty$ insures that Γ is rotund, since otherwise $\rho(s)$ is unbounded.

3. CONVEX CAUSTICS FOR BILLIARDS IN R^3

We would like to describe some of our recent progress in the search for convex caustics for billiards in R^3. ·

By a supporting line of a compact convex set K in R^3, we will mean a line L contained in a supporting plane of K such that $L \cap K \neq \phi$. Then Definition 1 above needs only a trivial modification to describe a convex caustic for billiards in R^3.

Our first guess in searching for caustics in R^3 was that the proper reflective container for a caustic would be defined by capbodies with constant *two*-dimensional surface area. Thus we investigated the following objects.

DEFINITION 5. Suppose that K is a compact convex set in R^3 with two dimensional surface area, $SA_2(K)$. Then the R^3-embedded profile of K determined by the constant λ, where $\lambda > SA_1(K)$, is $\{P \in R^3 \mid SA_2(\text{conv}(K \cup P)) = \lambda\}$. We will use γ_λ again to denote such profiles.

EXAMPLES. Unlike the planar case, there are no classical examples of R^3-embedded profiles, although they were studied by Stoll in [7, pp. 48-50, 55]. Let us briefly construct some elementary examples.

Let K be a singleton point. Then for all $P \in R^3$, conv(K $\cup$ P) is a closed line segment and $SA_2(\text{conv}(K \cup P)) = 0 = SA_2(K)$. Thus for any $\lambda > SA_2(K)$, $\gamma_\lambda = \phi$. Recall that R^2-embedded profile of K was a circle centered at K with positive radius. Thus profiles of K depend strongly on the choice of the embedding space.

Now let K be a line segment F_1F_2. If $P \in$ line F_1F_2, then conv(K $\cup$ P) is a closed line segment contained in this line, so $SA_2(\text{conv}(K \cup P)) = 0$ again. Hence the line F_1F_2 does not intersect any R^3-embedded profile of K. If $P \in R^3 \setminus$ line F_1F_2, thenconv(K $\cup$ P) is the relative interior of the triangle

PF_1F_2. Let $N(P)$ denote the orthogonal projection of $P - F_1$ on $F_2 - F_1$. Then $SA_2(\text{conv}(K \cup P)) = 2\left(\frac{1}{2} \|F_1 - F_2\| \cdot \|N(P)\|\right)$. (See Fig. 12.) Thus for $\lambda > SA_2(K) = 0$,

$$\gamma_\lambda = \{P \in R^3 \mid \|N(P)\| = \frac{\lambda}{\|F_1 - F_2\|} , \text{ a constant}\}$$

Hence γ_λ is a circular cylinder with line F_1F_2 as the central axis and radius equal to $\frac{\lambda}{\|F_1 - F_2\|}$. We note that in this case γ_λ is unbounded. Recall that the R^2-embedded profiles of segment F_1F_2 were ellipses.

Let K be the disk lying in the xy-plane centered at the origin with radius $a > 0$. Then $\text{conv}(K \cup P)$ is, in general, a solid oblique circular cone. Let $P = (x,y,z)$. Then it can be shown by using some calculus that

$$SA_2(\text{conv}(K \cup P)) = \frac{a}{2} \int_0^{2\pi} \sqrt{(a + \sqrt{x^2 + y^2} \cos \theta)^2 + z^2} \, d\theta + \pi a^2$$

Hence given $\lambda > 2\pi a^2 = SA_2(K)$,

$$\gamma_\lambda = \{(x,y,z) \mid \frac{a}{2} \int_0^{2\pi} \sqrt{(a + \sqrt{x^2 + y^2} \cos \theta)^2 + z^2} \, d\theta = \lambda - a^2\}$$

Thus γ_λ, in this case, is represented as the solution of an elliptic integral equation.

Finally, let K be a closed ball in R^3. If $P \in R^3 \setminus K$, then $\text{conv}(P \cup K)$ is a "dunce head," i.e., the convex hull of a right circular cone with vertex

FIGURE 12

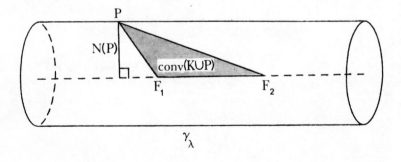

at P and sides tangent to K perched upon K. It should be clear that if $\lambda >$
$SA_2(K)$, then γ_λ will be a concentric sphere properly containing K. At last,
here is an example of an R^3-embedded profile of K which clearly exhibits the
reflective property of the R^2-embedded profiles; namely, the inner ball K is
a caustic for billiards inside its outer concentric ball, conv γ_λ.

From a billiards point of view these examples of R^3-embedded profiles
are disappointing. Except for the last example, it can be shown that K is
not a caustic for billiards inside conv γ_λ. Consider, for example, the line
segment $F_1 F_2$ once again with γ_λ a circular cylinder. The tangent plane to
$P \in \gamma_\lambda$ is a plane normal to the radius vector of the cylinder at P. Thus a
billiards trajectory which passes through F_1 and strikes γ_λ at a point P ra-
dially above F_2 is reflected away from segment $F_1 F_2$ towards the point $F_1' \in$
line $F_1 F_2$ which is symmetric to F_1 with respect to F_2. Thus this trajectory
travels along a supporting line of segment $F_1 F_2$ before reflection, but is not
reflected along a supporting line, since $PF_1' \cap K = \phi$.

A little reflection (no pun intended) should convince the reader that
the proper reflecting container for the line segment $F_1 F_2$ is a (prolate) ellipsoid of revolution using line $F_1 F_2$ as the axis of revolution for any ellipse focused at F_1, F_2. This example reveals that the search for caustics
in R^3 leads to consideration of nonprofiles in R^3. This conclusion is further buttressed by the following observation (suggested by a colleague, Burt
Cassler at LSU): if K is the disk lying in the xy-plane centered at the origin with radius a > 0, then K is a convex caustic for billiards in an (oblate)
ellipsoid of revolution. Choose $(-a,0,0)$ and $(a,0,0)$ as foci of an ellipse
in the xz-plane and revolve this ellipse about the z-axis. A rather long
computation using elliptic coordinates convinces one that the disk is indeed
a convex caustic for billiards inside the convex hull of such an ellipsoid.
The reader may have already noticed that every choice of K that we have used
in our examples above is a (degenerate) ellipsoid. Some further thought in
this direction led to the formulation of Theorem 3 below.

DEFINITION 6. A quadric surface S is said to be *confocal* to a given ellip-
soid $E = \{(x,y,z) \mid \frac{x^2}{a^2} + \frac{y^2}{b^2} + \frac{z^2}{c^2} = 1\}$ if there exists some $\lambda \in R$ such that

$$S = \{(x,y,z) \mid \frac{x^2}{a^2 \pm \lambda^2} + \frac{y^2}{b^2 \pm \lambda^2} + \frac{z^2}{c^2 \pm \lambda^2} = 1\}$$

REMARK. This definition is standard and may be found, for example, in Salmon [5, p. 166]. If we choose the positive sign in front of λ^2, then clearly S must be another ellipsoid. If we choose the negative sign in front of λ^2, however, there are three possibilities. If $\lambda^2 < \min\{a^2,b^2,c^2\}$, then S is still an ellipsoid. Supposing that $c^2 = \min\{a^2,b^2,c^2\}$, if $c^2 < \lambda^2 < \min\{a^2,b^2\}$, then S is a hyperboloid of one sheet. Finally, supposing that $b^2 = \min\{a^2,b^2\}$, then S is a hyperboloid of two sheets. Thus the system of quadrics confocal to a given ellipsoid includes a one parameter family of ellipsoids, a one-parameter family of hyperboloids of one sheet, and a one-parameter family of hyperboloid of two sheets. Such a system of quadrics enjoys many beautiful geometric properties and the interested reader is urged to consult the well-illustrated book by Hilbert and Conn-Vossen [2, pp. 19-25] for an introductory account, as well as Salmon [5, pp. 151-201]. We will use one well-known property of these confocal systems below.

THEOREM. The axes of any tangent cone to a quadric are the normals to the three confocals which can be drawn through the vertex of the cone.

   Proof. The reader is referred to Salmon [5, pp. 173-174] for a proof.

REMARK. We are now prepared to state Theorem 3 which encompasses all the examples of convex caustics for billiards in R^3 discussed above, provided the proper interpretation of a confocal system is made for the degenerate ellipsoids.

THEOREM 3. Suppose E is any ellipsoid in R^3. Then $K = \text{conv } E$ is a convex caustic for billiards inside conv E^1 provided that E^1 is a confocal ellipsoid to E and conv $E \subseteq \text{conv } E^1$.

   Proof. Consider any billiards trajectory travelling along a supporting line to conv E before reflection at $P \in E^1$. Since a supporting line to conv E is a tangent line to E, the trajectory travels along a generator of the tangent cone C to E with vertex at P. Let us call this generator the incident generator. The tangent plane to E^1 at P is determined by the normals to the confocal hyperboloids to E which pass through P. Thus by the theorem in Salmon cited above, the central axis of the cone C is the normal line N to E^1 at P. If we look down this normal line N from P, E appears as a flat ellipse center at $N \cap E$, as in Fig. 13. Let the plane determined by the incident

FIGURE 13

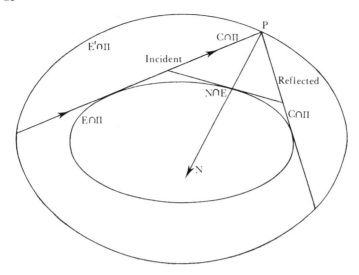

generator and N be denoted by Π. Let the generator of the tangent cone C to E which lies in the plane section $C \cap \Pi$ and is not the incident generator be called the reflected generator. We call this generator the reflected generator since N, beingthe central axis of a right elliptical cone as seen from P, bisects the angle between the incident generator and the reflected generator. Thus the billiards trajectory after reflection at P travels along the reflected generator of C, which is also a supporting line of E. This shows that conv E is a convex caustic for billiards inside conv E^1. $\Box$

4. CONCLUSION

It can be shown that if E and E^1 are confocal ellipsoids as above, that $SA_2(\text{conv}(E \cup P))$ where $P \in E^1$ is not a constant function. Thus the confocal ellipsoids of E whose convex hulls properly contain E are not R^3-embedded profiles of conv E. This indicates rather conclusively that two-dimensional surface area is not the metric invariant we need to use in R^3 to find the reflective container in which a given compact convex set K is a convex caustic for billiards. It is interesting to note in this regard that the system of confocal ellipsoids to a given ellipsoid may be generated by a string construction due to Staude (cf. Hilbert and Conn-Vossen [2, pp. 19-20]). Thus it is possible that this construction may be generalized to convex sets which

are not quadrics. We hope to be able to report on this prospect in the near
future.

ACKNOWLEDGMENTS

It is a pleasure to express gratitude to Professor Walter Gottschalk
of Wesleyan University for his patient and inspiring supervision of this research. The author would also like to thank his colleagues Richard Anderson,
Burt Casler, Doug Curtis, and Jan van Mill for their encouragement and support.

REFERENCES

1. T. Bonnesen and Fenchel, W., *Theorie der konvexen Körper*, Chelsea, New
    York, 1948.

2. D. Hilbert and Conn-Vossen, S., *Geometry and the Imagination* (trans:
    P. Nemenyi), Chelsea, New York, 1952.

3. V. F. Lazutkin, The existence of caustics for billiards in a convex do-
    main, *Math. USSR Izvestia* 7(1973), 185-214.

4. E. H. Lockwood, *A Book of Curves*, Cambridge University Press, Cambridge,
    1961.

5. G. Salmon, *Analytic Geometry of Three Dimensions*, Vol. I, Longmans,
    Green and Co., 1914.

6. Ya. G. Sinai, *Introduction to Ergodic Theory*, (trans: V. Scheffer),
    Princeton University Press, Princeton, 1976.

7. A. Stoll, Ueber den Kappenkörper eines konvexen Körpers, *Comment. Math.
    Helv.*, 2(1930), 35-68.

8. P. H. Turner, On the smoothness of profiles, *Notices of the AMS*, 26(1979),
    Abstract 79T-D8.

9. P. H. Turner, Caustics characterize R^2-embedded profiles, *Abstracts of
    the AMS*, 1(1980), Abstract 773-78-1.

10. P. H. Turner, *Aspects of Convexity in Billiard Ball Dynamical Systems*,
    Doctoral Thesis (1980), Wesleyan University, Middletown, CT.

APPENDIX: A BIBLIOGRAPHY FOR BILLIARDS

Arnold, V. I., Small denominators and problems of stability of motion in
    classical and celestial mechanics, *Uspehi Mat Nauk* 18(1963), 183-184.

Arnold, V. I. and Avex, A., *Ergodic Problems of Classical Mechanics* (trans:
    A. Avez), W. A. Benjamin, New York, 1968, 230-234.

Birkhoff, G. D.. *Dynamical Systems* (Revised Edition), AMS Colloquium Publications, Vol. IX, Providence, 1966, 169-179.

Birkhoff, G. D., On the periodic motions of dynamical systems, *Acta Mathematica* 50(1927), 359-379.

Birkhoff, G. D., Some unsolved problems of theoretical dynamics, *Science* 94 (1941), 599.

Birkhoff, G. D., What is the ergodic theorem? *Amer. Math. Mo.* 49(1942), 222-226.

Birkhoff, G. D., El problema de la bola de billar y su significado en la dinamica moderna, *Revista de Ciencias* 44(1942), 244-245.

Boldrighim, C.,; Keane, M.; and Marchetti, F., Billiards in polygons, *Ann. Probab.* 6(1978), 532-540.

Croft, H. T.; Swinnerton-Dyer, H. P. F., On the Steinhaus billiard table problem, *Proc. Camb. Phil. Soc.* 59(1963), 37-41.

Dvorin, M. M. and Lazutkin, V. F., The existence of an infinite number of elliptical and hyperbolic periodic trajectories for a convex billiard, *Functional Analysis and Applications* 7(1973), 20-27.

Fox, R. H. and Kershner, R. B., Concerning the transitive properties of geodesics on a rational polyhedron, *Duke Math. Journal* 2(1936), 147-150.

Gallavotti, G., Lectures on the Billiard, *Dynamical Systems, Theory and Applications* (ed. J. Moser), Springer-Verlag, Berlin, 1975.

Halpern, B., Strange billiard tables, *Trans. Amer. Math. Soc.* 232(1977), 297-305.

Hardy and Wright, *An Introduction to the Theory of Numbers*, 3rd ed., Clarendon Press, Oxford, 1954, 376-379.

Klee, V., Is every polygonal region illuminable from some point? *Amer. Math. Monthly* 76(1969), 180.

Klee, V., Some unsolved problems in plane geometry, *Math. Mag.* 52(1979), 134-135, 141.

Konig, D. and Szucs, A., Mouvement d'un point abandonné a l'intérieur d'un cube, *Rendi. Circ. Mat. di Parlermo* 36(1913), 79-90.

Lazutkin, V. F., On the asymptotics of the eigenfunctions of the Laplacian, *Soviet Math. Dokl.* 12(1971), 1569-1572.

Lazutkin, V. F., The existence of a continuum of closed invariant curves for a convex billiard, *Uspehi Mat. Nauk* 27(1972), 201-202.

Lazutkin, V. F., The existence of caustics for billiards in a convex domain, *Math USSR Izvestia* 7(1973), 185-214.

Morse, M., What is analysis in the large? *Studies in Global Geometry and Analysis* (ed. S. S. Chern), MAA and Prentice Hall Inc., 1967, 12-13.

Poritsky, H., The billiard ball problem on a table with a convex boundary--
an illustrative dynamical problem, *Ann. Math.* 51(1950), 446-470.

Rauch, J., Illumination of bounded domains, *Amer. Math. Mo.* 85(1978), 359-361.

Pastor, J. Rey, The last geometric theorems of Poincaré and their applications, *Union Mat. Argentina* 1(1945), 42; *Math. Rev.* 7(1946), 471.

Santaló, L., *Integral Geometry and Geometric Probability*, Addison-Wesley,
Reading, 1976, 39-41, 60-62.

Schoenberg, I. J., On the motion of a billiard ball in two dimensions, *Delta*
5(1975), 1-18.

Schoenberg, I. J., Extremum problems for the motions of a billiard ball, I.
The L_p norm $1 < p < \infty$, *Indag. Math.* 38(1976), 66-75.

Schoenberg, I. J., Extremum problems for the motions of a billiard ball, II.
The L_∞ norm, *Indag. Math.* 38(1976), 263-279.

Sinai, Ya. G., Dynamical systems with elastic reflections: Ergodic properties
of dispersing billiards, *Uspehi Mat. Nauk* 25(1970), 141-192.

Sinai, Ya. G., *Introduction to Ergodic Theory* (trans. V. Scheffer), Princeton University Press Mathematical Notes No. 18, Princeton, 1976, 81-90,
143-150.

Sinai, Ya. G., Billiard trajectories in a polyhedral angle, *Uspehi Mat. Nauk*
33(1978), 229-230.

Sine, R., A characterization of the ball in R^3, *Amer. Math. Monthly* 83(1976),
260-261.

Sine, R., and Kreinovic, V., Remarks on billiards, *Amer. Math. Monthly* 86
(1979), 204-206.

Sternberg, S., *Celestial Mechanics*, Vol. II, W. A. Benjamin, New York, 1969,
199-205.

Steinhaus, H., *Mathematical Snapshots*, 3rd ed., Oxford University Press, New
York, 1969.

Sudan, G., Sur le probleme du rayon réfléchi, *Revue Roumaine de Math. Pures
et Appliquées* X(1965), 723-733.

Turner, P. H., Convex caustics for billiards in R^2 and R^3 (these *Proceedings*,
1981).

Zemlyakov, A. N. and Katok, A. B., Topological transitivity of billiards in
polygons, *Mat. Zametki* 18(1975), 291-300.

ON PACKING CURVES INTO CIRCLES

Erwin Lutwak

Department of Mathematics
Polytechnic Institute of New York
Brooklyn, New York

The letter K, possibly with subscripts, will be used exclusively to denote a compact convex set in the Euclidean plane. The length of ∂K will be noted by $L(K)$. The function δ_K is defined on $\partial K \times \partial K$ by letting $\delta_K(x,y)$ denote the distance between the points x and y. The function ζ_K is defined on ∂K by letting

$$\zeta_K(x) = \max_{y} \delta_K(x,y)$$

where the maximum is, of course, taken over all y in ∂K. Alexander [1] conjectured that

$$\min_{x} \zeta_K(x) \leq L(K)/\pi \tag{1}$$

with equality if and only if K is of constant width.

Falconer [2], using Poincaré's formula [9], recently proved the conjecture of Alexander.

The compact convex set of points lying on one side of a diameter of a disc of radius r will be called a *semidisc* of radius r. The following result was proved by A. Meir in answer to a question of L. Moser.

Every plane curve of unit-length is contained in some semidisc of radius 1/2.

The proof given by Meir is reproduced in [10]. The following analogous result for closed curves is an immediate consequence of (1).

THEOREM 1. Every closed plane curve C of unit length is contained in some semidisc of radius $1/\pi$; in fact, unless C bounds a convex set of constant width, C is contained in some semidisc of radius less than $1/\pi$.

To prove this, let K be the convex hull of C. It is well known that L(K) is not greater than the length of C. If we let c be a point of K at which ζ_K attains a minimum, it follows from (1) that $\zeta_K(c) \leqslant 1/\pi$, and $\zeta_K(c) = 1/\pi$ if and only if K is of constant width. If ℓ is a support line of K through the point c, then K will be contained in a semidisc of radius $\zeta_K(c)$ whose diameter lies along ℓ and whose "center" is c. $\square$

A circle C is said to *accommodate* the convex sets $K_1, \ldots, K_n$ if there exist rigid motions $\phi_1, \ldots, \phi_n$ such that the interiors of the sets $\phi_i(K_i)$ are pairwise disjoint and contained in the region bounded by C. The following theorem is a simple consequence of Theorem 1.

THEOREM 2. The smallest circle C that will accommodate two discs of unit length will also accommodate any two compact convex sets of unit length (which, besides having equal length, need not be related); in fact, unless both convex sets are of constant width, they can be accommodated in a circle smaller than C.

In light of Theorem 1, the only thing we need verify in order to prove Theorem 2 is that if K_1 and K_2 are of unit length, and K_1 is of constant width, while K_2 is not of constant width, then K_1 and K_2 can be accommodated in a circle smaller than C. To do this we require a preliminary definition.

If S is a semidisc of radius r and ℓ is the line perpendicular to the diameter of S that passes through the "center" of S, then the compact convex set of points lying in S which are a distance no greater than $(r\sqrt{3})/2$

from ℓ will be called a *clipped semidisc* of radius r and will be denoted by S'.

Since K_1 is of unit length and constant width it must be contained in a clipped semidisc of radius $1/\pi$. To see this, let x, y be the endpoints of a diameter (longest chord) d of K. It is well known that the diameters of a unit length convex set of constant width must all have length $1/\pi$. The line ℓ which is perpendicular to d and passes through x is clearly a support line of K_1. Let S be the semidisc of radius $1/\pi$ "centered" at x, whose diameter lies along ℓ and which contains y. Clearly, K is contained in S. Since no point of K can be a distance greater than $1/\pi$ from either x or y, K must, in fact, be contained in S'.

From Theorem 1 it follows that since K_2 is not of constant width it must be contained in a semidisc of radius less than $1/\pi$. It is easy to see that a clipped semidisc of radius $1/\pi$ and a semidisc of radius of less than $1/\pi$ can be accommodated in a circle whose radius is less than $1/\pi$. Thus, K_1 and K_2 can be accommodated in a circle whose radius is less than $1/\pi$, as required. $\square$

A circle C is said to accommodate the simple closed plane rectifiable curves $C_1,\ldots C_n$ if there exist rigid motions $\phi_1,\ldots,\phi_n$ such that the interiors of the regions bounded by the $\phi_i(C_i)$ are pairwise disjoint and contained in the region bounded by C. Obviously, Theorem 2 could be restated as:

THEOREM 2'. The smallest circle that will accommodate two circles of unit length will also accommodate any two simple closed plane curves of unit length; in fact, unless both curves bound convex sets of constant width, they can be accommodated in a circle smaller than C.

It should be noted that while we have restricted our attention to plane rectifiable curves that are simple, Theorem 2' can easily be extended to curves that are not necessarily simple.

From Theorem 2 one is led to the following question:

Will a circle that accommodates the discs D_1 and D_2, necessarily accommodate the convex sets K_1 and K_2 that have the property $L(K_1) = L(D_1)$ and $L(K_2) = L(D_2)$?

If D_1 and D_2 are the same length, then an affirmative answer to the question is provided by Theorem 2; but, in general, the answer is no. In fact, it is possible that a circle C may accommodate D_1 and D_2 and, yet, not accommodate even K_1 by itself. To see this, consider the case where D_1 is a much larger disc than D_2. In such a case, a circle C that is only slightly larger than the boundary of D_1 could accommodate both D_1 and D_2. Now if K_1 is a "long thin" convex set whose diameter is close to $L(K_1)/2$, then C could not accommodate even K_1 by itself.

Another question that arises is:

If C is the smallest circle that will accommodate n unit-length discs, will C necessarily accommodate any n unit-length compact convex sets?

Theorem 2 is, of course, the case n = 2, and it seems likely that the answer to the question is affirmative for at least some values of n other than 2.

It should be noted that the problem of determining the radius of the smallest circle that will accommodate n unit-length discs is solved only for a vew small values of n (see [3, 4, 5, 6, 7, 8]).

Theorem 2 also leads to questions regarding the validity of its extensions to higher dimensional spaces. If we attempt to extend Theorem 2 to higher dimensional spaces by simply replacing "length" by "surface area" in the statement of Theorem 2, we obtain a false statement. In fact, the sphere of smallest radius that will accommodate two balls of unit surface area need not even accommodate one convex set of unit surface area; specifically, it would not accommodate a "long thin needle" of unit surface area.

An extension of Theorem 2 to higher dimensional spaces would be provided by an affirmative answer to the following question:

If a sphere accommodates two balls of unit mean width, will it necessarily accommodate any two compact convex sets of unit mean width?

REFERENCES

1. R. Alexander. Metric embedding techniques applied to geometric inequalities, in *The Geometry of Metric and Linear Spaces* (L. M. Kelly, ed.), Springer-Verlag, Berlin, 1975, 57-65.

2. K. J. Falconer. A characterization of plane curves of constant width, *J. London Math. Soc.* (2) 16(1977), 536-538.

3. M. Goldberg. Packing of 14, 16, 17 and 20 circles in a circle, *Math. Mag.* 44(1971), 134-139.

4. L. Fejes Tóth. *Lagerungen in der Ebene, auf der Kugel und in Raum*, Springer-Verlag, Berlin, 1953.

5. S. Kravitz, Packing cylinders into cylindrical containers, *Math. Mag.* 40(1967), 65-71.

6. U. Pirl, Der Mindestabstandt von n in der Einheitskreisscheibe gelegenen Punkten, *Math. Nach.* 40(1969), 111-124.

7. G. E. Reis, Dense packing of equal circles within a circle, *Math. Mag.* 48(1975), 33-37.

8. C. A. Rogers, *Packing and Covering*, Camb. Univ. Press, New York, 1964.

9. L. A. Santaló, *Integral Geometry and Geometric Probability*, Addison-Wesley, Reading, Mass., 1976.

10. J. E. Wetzel, Sectorial covers for curves of constant length, *Canad. Math. Bull.* 16(1973), 367-375.

A PERSPECTIVE ON ABSTRACT CONVEXITY: CLASSIFYING ALIGNMENTS BY VARIETIES

Robert E. Jamison-Waldner

Department of Mathematical Sciences
Clemson University
Clemson, South Carolina

1. INTRODUCTION

The primary goal of this presentation is to report on an attempt to formulate a suitable classification scheme for *alignments*--certain systems of abstract "convex" sets. But since the area of "abstract convexity" is at best only loosely established, there are certain questions which call out for answers before such a report can even be begun. Why abstract convexity? What are alignments? How do they fit in with the rest of mathematics? Why should they be classified as proposed here rather than in some other way? Besides providing motivation for the proposed classification scheme, answers to these questions entail important connections between convexity and other branches of mathematics, which up to now seem to have been largely overlooked by workers in axiomatic convexity.

Therefore, we have chosen to sacrifice the theorem-proof regalia in favor of a more informal discussion aimed at providing some answers to the above questions as well as a glimpse of the spirit of the classification scheme. The technical details and omitted theorems will perhaps someday find their way into a separate monograph.

In the beginning of his *Introduction to a Form of General Analysis* [45], E. H. Moore laid down his principle of generalization: the existence of analogies between various specific theories implies the existence of a general theory which unifies these analogies on the basis of a common underlying structure. Following this guideline, the goal of the theory of alignments is not so much to "generalize Euclidean convexity" as to unify the geometric aspects of a variety of structures, including vector spaces, modules, ideals, graphs, matroids, ordered sets, lattices, and semilattices. (Several illustrations of analogies are provided in Section 2.) Euclidean convexity is just one piece in the mosaic.

On the other hand, the theory of alignments does not attempt to embrace all possible uses of the term "convex." Convexity is, after all, a very broad geometric concept. An all-inclusive theory would have to be so general as to be practically useless. Otherwise, it would place undesirable restraints on the imaginative use of "convexity" as a basic, intuitive concept.

Alignments, formally defined in Section 3, are *algebraic closure systems* [2, 6, 19], the "closed" sets associated with finitary hull operators. As such they capture only the basic combinatorial and finitary aspects of a system of convex sets and their associated hull operator. More general systems of sets could be studied as abstractions of convexity but, I believe, with less success. The goal is not extreme generality but rather unity of approach. Alignments have significant links with universal algebra, lattice theory, and the first order logic of finitary relational systems which will be touched on in Section 6. Although our basic view of alignments will be as geometric "convexity spaces" endowed with a hull operator, it is always well to keep in mind the algebraic background of these structures.

The study of abstract hull operators, although often put under the label "axiomatic convexity" by convex geometers, historically has several valid avenues of approach, convexity being only one. The long tradition apparently has its beginning in E. H. Moore's 1905 book *Introduction to a Form of General Analysis* (sections 28-42 of [45]). The early development includes Kuratowski's celebrated closure axioms for a topology, as well as work motivated by ideal theory [50, 51], logic [54], and lattice theory [3]. Valuable surveys are given by Jürgen Schmidt in [50] and [51]. (The latter is unfortunately not widely available.)

Convexity is a relative latecomer to the abstract theory of hull operators. The connection, however, is clear since the convex hull operator
plays such a fundamental role in convexity theory. Stimulated by the survey of Danzer, Grünbaum, and Klee [9], much of the work [14, 15, 23, 30, 36,
41, 48, 52, 53] (see also Sierksma's chapter in these *Proceedings*) on abstract
convex hulls has sought a better understanding of the classical theorems
of Carathéodory, Helly, and Radon. Unfortunately, the significance of the
finitary axiom (that the hull operator is determined in general by its
action on finite sets) was missed by many of these authors, largely on the
grounds that it was not needed and more generality could be obtained by
omitting it. In fact, this is shallow generality. It can be easily shown
(see pages 8 and 9 of [29]) that any hull operator and the alignment it
generates *agree* on all *finite* sets. Thus, in combinatorial studies (as
those around the three classical invariants) which involve only hulls of
finite sets, one can assume without loss of generality that one has an
alignment. The finitary axiom is actually an invaluable tool for lifting
results from the finite to the general case.

The important distinction between *algebraic* (finitary) and *topological*
closure operators dates back at least to the 1940's (see p. 300 of [3]).
Nonetheless, feelings have persisted that all hull operators should be
viewed as "generalized" topologies. Given the nearly universal familiarity
with topology by mathematicians, it is not difficult to see how this view
could arise. In 1953 Jürgen Schmidt already warned (page 41 of [51]) of
the dangers of this view:

> Such an undertaking runs the risk of interpreting all
> concepts of general hull-theory topologically, and thereby
> restricting its development to those concepts and questions
> which have proven meaningful in the preceding development
> of topology.

Topologies and alignments really represent two different aspects of
geometry. If topology is the *rubber* sheet geometry, in which structures
can be stretched and deformed because only limiting behavior is important,
then alignments are *rigid* sheet geometries, basically combinatorial in
nature. There is, of course, an important overlap of these two types of
geometry. Many general closure operators can be viewed as "closed convex
hull" operators for appropriate choices of alignment and topology on the
underlying space. That is always the case, for example, if the closure
system consists of closed subsets of a compact space and satisfies the
Blaschke Selection Theorem [29, p. 50].

Thus, in this report, an effort is made to stress the importance of
the finitary property and avoid forcing alignments into the mold of topology.
Following Moore's principle of generalization, the approach is eclectic,
borrowing examples and intuition from a variety of disciplines. This is
reflected in the terminology. "Alignment" is used in preference to "domain finite convexity structure" [22, 52, 53], "algebraic closure system"
[6, 19], and "geometry" [65] to emphasize that alignments are a class of
structures worthy of study in their own right. It suggests something
geometric and rigid without evoking an established viewpoint. Moreover,
it is short; it inflects well; it is not the name of an intuitive mathematical concept, and it is almost unused elsewhere in mathematics. (There
are "aligned functionals" in optimization.)

The sets in an alignment will be called "convex" rather than "closed"
and the separation axioms will be labelled S_0, S_1, S_2, S_3, S_4, analogous
but not identical to the topological "Trennungsaxiome." This also avoids
such awkward and ambiguous phrases as "a closed closed set in a T2 T3
space" in the emerging area of topological alignments [28, 29, 31, 60].
Other borrowed terminology includes "irredundant" (lattices), "valence"
(graphs), "copoint" (matroids), and "variety" (universal algebra).

It may seem peculiar to use the term "convex" in some of the instances
where alignments arise, say, for the ideals in a ring. In his development
[50, 51], Jürgen Schmidt, as an algebraist, chose to call the members of
an arbitrary alignment "ideals," and he seems to have viewed the convex
sets in the plane as an anomolous example (see page 38 of [51]). As a
geometer, this author is more comfortable with the geometric terminology.
In any case, any analogy invoked by terminology is intended only as a guide
and not as dogma.

2. WHY ABSTRACT CONVEXITY?

One possible reason to abstract convexity is that it can be done so
easily. It is not at all hard to invent a reasonable set of axioms for
a system of convex sets, a convex hull operator, or a generalized segment
function. But this can hardly be considered a satisfactory answer to the
title question.

Moore's "principle of generalization" demands more. It demands
analogies which can be unified using concepts from convexity. There are
in fact many of these. By way of illustration, we have chosen four examples,

three of which deal with the "classical invariants" of Helly, Radon, and Carathéodory. (See [9, 36, 41, 52] for definitions.)

A. The Chinese Remainder Theorem as a Helly-type Theorem

One version (page 279 of [67]) of the Chinese Remainder Theorem is this: Given a finite system of congruence equations

$$x \equiv a_i \mod m_i$$

in the integers Z, the system has a solution provided each pair of congruences has a solution. Suppose we call a set K of integers "convex" provided it is a coset (translate) of some ideal in Z. Then K is "convex" if and only if it has the form

$$K = (z \in Z: \quad z \equiv a \mod m)$$

for some modulus m and some residue a. The above form of the Chinese Remainder Theorem then asserts that a finite family of "convex" sets in Z has nonvoid intersection if each *two* of them have nonvoid intersection. That is, it asserts that the Helly number of this "convexity" on Z is 2.

For a further discussion of this form of the Chinese Remainder Theorem, see pages 221 and 265 of [19] or Wille [65].

B. Radon's Theorem and Equal Unions of Sets

Lindstrøm [44] has shown that if $\Re$ is a family of n + 1 or more subsets of an n-element set X, then there are two disjoint subfamilies $\Re_1$ and $\Re_2$ of $\Re$ whose set-theoretic unions are equal. Suppose we define a family $\mathcal{L}$ of subsets of X to be "convex" provided $\mathcal{L}$ is closed under unions: A and B in $\mathcal{L}$ implies A $\cup$ B in $\mathcal{L}$. Then the "convex hull" of any family is just the family of all possible unions of sets from that family. Thus, Lindstrøm's theorem is equivalent to the assertion that if $\Re$ is a family of n + 1 or more sets, then $\Re$ can be partitioned into two subfamilies whose "convex hulls" have nonempty intersection. That is, n + 1 is the Radon number of this notion of convexity on the power set Pow(X) of X.

Tverberg [59], in fact, gave an ingenious proof of Lindstrøm's result using Radon's Theorem in Euclidean space. For a further discussion, see [39].

C. Hyperplane Coverings Over Finite Fields and Carathéodory's Theorem

If V is a vector space over a field F, a subset A of V is called *affine* if it is a translate (coset) of a vector subspace of V. As is easily observed, every affine set is an intersection of hyperplanes. A set is called *coaffine* [34, 35] if it is the intersection of *complements* of hyperplanes.

Suppose now that F is a finite field and that V is finite dimensional over F. A covering of V by hyperplanes through O is *reduced* if no sub-family of the cover also covers V. Let N(V,F) be the *maximum number* of hyperplanes through O in a reduced cover of V. If we interpret "convex" in V to mean coaffine, then we have this result [34]:

(2.1) Any point in the coaffine hull of a set S is in the coaffine hull of a subset of S of at most N(V,F) points.

This is an analog of Carathéodory's Theorem. It can be shown [34] that for fixed F, N(V,F) grows polynomially in the dimension of V whereas the cardinality $|V|$ grows exponentially. Thus, (2.1) is a nontrivial result. There is a similar relationship between a hyperplane covering number and the Helly number of the coaffine sets [34].

D. Chordal Graphs and the Krein-Milman Property

If G is any simple graph (undirected without loops or multiple edges), a set K of the nodes of G is *geodesically convex* if for each pair of nodes u and v in K, *all* nodes on *all* shortest paths from u to v are also in K. Although this is a natural definition [25] of convexity in graphs, it is not always the most useful. A path P in G is *chordless* if the only pairs of edges in P that are adjacent in G are consecutive along P. A set K of nodes is *monophonically convex* if for any u and v in K, *all* nodes on *all* chordless paths from u to v are also in K. Since shortest paths are always chordless, this is a stronger requirement than for geodesic convexity.

A graph G is said to be a *chordal* ("rigid circuit" or "triangulated") graph if each cycle of four or more nodes in G has a chord (i.e., an edge in G joining two nonconsecutive nodes of the cycle). Such graphs arise in the famous Lekkerkerker-Boland characterization of interval graphs [43] as well as in other contexts [1, 13]. A node v of a graph G is *simplicial* (or "complete") if every pair of neighbors of v are adjacent in G.

It is easy to see that the simplicial nodes are precisely the *extreme-points* of G--with regard to either geodesic or monophonic convexity [37]. (We define an extremepoint p of a convex set K as a point p such that K \ p is still convex.)

The following result [37] states that chordal graphs are precisely those with the Krein-Milman property.

(2.2) For any graph G, the following are equivalent:

   (i)   G is chordal;

  (ii)   any monophonically convex set in G is the monophonic convex hull of its extremepoints;

 (iii)   for any node v in G and any radius $r > 0$ the ball $B_r(v)$ of all nodes joined to v by a path of length r or less is monophonically convex.

Condition (ii) improves a result of Dirac [13] that every chordal graph has at least one simplicial vertex. Indeed, since single nodes are monophonically convex, if a graph of more than one node is to be the hull of its simplicial vertices, it must have at least two of them.

Condition (iii) implies that the center of a chordal graph G is connected--generalizing the familiar result that the center of a tree is either a single node or a pair of adjacent nodes. Indeed, if r is the radius of G, then it follows at once from the definitions that the center C of G is given by

$$C = \bigcap_{v \in G} B_r(v)$$

By (iii) the balls $B_r(v)$ are convex, so their intersection is also convex. But the definition of convexity implies that any convex set is connected since it contains all shortest paths.

Although these corollaries can be derived without the use of convexity, they do show that convexity can play a natural role in graph theory.

There are other illustrations of convexity-like ideas outside the realm of ordinary convexity. A theorem of Doignon [14] on Helly numbers for convexity on lattice points was rediscovered by Scarf [49] in an investigation of constraints in integer programming. Richard Wilson [66] introduced a finitary hull operator on the integers in his study of existence of block designs. Van Maaren [61, 62] and De Smet [11] have studied notions of convexity over general fields. There are analogs of the theorem of Tietze on locally convex continua for semilattices and lattices [31].

    All of these indicate that there are indeed legitimate reasons for abstract convexity.

3. ALIGNMENTS AND VARIETIES: DEFINITIONS

    The basic structures in our abstraction of convexity all consist of a set X together with some family $\mathcal{L}$ of distinguished subsets of X which are regarded as *abstract convex subsets* of X. The convex sets will always be required to satisfy the following axioms:

A1(a). ϕ is convex;

A1(b).   X is convex;

A2.   the arbitrary intersection of convex sets is convex;

A3.   the union of any family of convex sets totally ordered by inclusion is again convex.

Such a family $\mathcal{L}$ of abstract convex sets will be called an *alignment* on X, and the pair $(X, \mathcal{L})$ will be called an *aligned space*. Note that axiom A3 is trivially satisfied if X is finite. But it is *not* unimportant; on the contrary, it is the vehicle allowing one to reduce many problems on alignments to the finite case.

    Axioms A1(b) and A2 insure that for any subset S of X there is a smallest convex set $\mathcal{L}(S)$ containing S--the *hull* of S. The hull operator possesses the following properties:

H1. $\mathcal{L}(\phi) = \phi$

H2. $S \subseteq \mathcal{L}(S)$

H3. $S \subseteq T$ implies $\mathcal{L}(S) \subseteq \mathcal{L}(T)$

H4. $\mathcal{L}(\mathcal{L}(S)) = \mathcal{L}(S)$

H5. $\mathcal{L}(\cup \{S_i : i \in I\}) = \cup \{\mathcal{L}(S_i) : i \in I\}$ for any chain of sets S_i $(i \in I)$.

From the relations

$$\mathcal{L}(S) = \cap \{L \in \mathcal{L} : S \subseteq L\}$$

and

$$\mathcal{L} = \{L \subseteq X: \quad L = \mathcal{L}(L)\}$$

it is clear that an alignment and its hull operator uniquely determine each other. From axiom H5 (equivalent to A3), one can deduce the important fact that the hull operator associated with an alignment is finitary [2, 6, 22, 29, 51]:

(3.1) FINITARY PROPERTY: For any subset S of X and p in X, if $p \in \mathcal{L}(S)$, then $p \in \mathcal{L}(E)$ for some finite subset E of S.

It is clear from this that *polytopes*--the convex hulls of finite sets-- play a special role (cf. Section 6.B).

Also central to the development is the notion of a *subspace*. If Y is a subset of X, then any alignment $\mathcal{L}$ on X can be *restricted* to Y in a natural way:

$$\mathcal{L}|Y = \{L \cap Y: \quad L \in \mathcal{L}\}$$

This is then an alignment on Y whose hull operator is given by

$$\mathcal{L}|Y: \quad S \rightarrow \mathcal{L}(S) \cap Y$$

We can now express the finitary property as follows (cf. Section 5.B):

(3.2) Every aligned space is uniquely determined by its finite subspaces.

This is the key to our definition of varieties.

By a *variety of alignments*, we shall mean a class V of aligned spaces with the following properties:

    V1. Any space isomorphic to a space in V is also in V,

    V2. Any subspace of a space in V is also in V,

    V3. If every finite subspace of a space $(X,\mathcal{L})$ is in V, then $(X,\mathcal{L})$ is in V.

By an *isomorphism*, we mean, of course, a one-to-one correspondence pre- serving convexity both ways.

The variety consisting only of the 1-point space will be denoted ⟨1 pt⟩ and called the *trivial variety*. The class of all aligned spaces is the *universal variety* denoted by ⟨all⟩. In general, varieties will be denoted by pointed brackets around a short mnemonic description—e.g., ⟨downset⟩, ⟨TO⟩, ⟨rank ⩽ n⟩. (These terms are defined in the next section.)

The term "variety" is used here in the sense of a "natural class of similar spaces," suggesting a spiritual rather than a formal relationship with the varieties of universal algebra and algebraic geometry. The definition of "variety" is based on the empirical observation that many natural classes—e.g., totally ordered sets, partially ordered sets, and matroids—of alignments as well as many classes defined by natural conditions—e.g., separation axioms, a finite Krein-Milman property, bounds on Carathéodory number and rank—share two common features. First, the class is closed under the formation of subspaces (axiom V2), a very natural construction from the geometric point of view. Second, the class is of *finite character* (axiom V3), a very natural property from the algebraic point of view. Thus, the definition of variety embodies the spirit of the geometric-algebraic duality of alignments.

The definition of a variety is ready-made to allow a classification of alignments by forbidden substructures. This type of characterization is currently enjoying considerable popularity in combinatorics. (See for example [55].) Its distinguished history includes Kuratowski's celebrated characterization of planar graphs [24], the Lekkerkerker-Boland characterization of interval graphs [43], and Tutte's work on regular and graphic matroids [56-58]. By axiom V3, if a space X is not in a variety V, then some *finite* subspace of X fails to be in V, and hence there must be a smallest subspace F of X not in V. Such a subspace F must be a *minimal forbidden space* (MFS) for V in the sense that F is not in V but every proper subspace of F is in V. Clearly then if one knows all minimal forbidden subspaces for V, then one has a characterization of V. We shall discuss this idea in conjunction with concrete examples in Section 7.

It is also clear from axiom V3 that every variety is completely determined by the finite spaces it does contain. Suppose we choose for each isomorphism class of finite aligned spaces one representative example and denote the set of choices by *FinALN*. Then any variety can be identified by its intersection with FinALN, which is a countable, graded partially ordered set. It is countable since there are at most 2^{2^n} alignments on an n element set; it is graded by cardinality, and it is partially ordered by saying $Y \leqslant X$ iff Y is isomorphic to a subspace of X. Varieties are then in 1-1 correspondence with the subsets Δ of FinALN with the property: $X \in \Delta$ and $Y \leqslant X$ implies $Y \in \Delta$.

FinALN provides us with a convenient way of thinking of varieties. For a purist, it may relieve the anxiety to state that, due to axiom V1, varieties are really "too big" to be sets. The minimal forbidden subspaces for a variety correspond to the minimal elements of its complement in FinALN. As such the characterization by forbidden subspaces is just an illustration of the familiar correspondence between downsets (order ideals) and antichains in graded ordered sets. Needless to say, it would be very nice to have more order-theoretic information about FinALN.

4. EXAMPLES OF ALIGNMENTS AND VARIETIES

In section 2 we noted several notions of convexity which defined alignments on various structures. Here we will augment that list with other important examples. It is well worth noting that a given structure often carries more than one natural alignment. For example, a real vector space carries both the *ordinary alignment* of all usual convex sets and the *affine alignment* of all affine flats. Vector spaces over finite fields carry both the affine and coaffine alignments. (The coaffine sets form an alignment iff the underlying field is finite [35].) The node set of a graph is endowed with both the geodesic and the monophonic alignments.

A. Ordered Sets

Outside of real vector spaces, the term "convex set" is probably most often applied in the context of (partially) ordered sets [46]. A subset K of an ordered set X is *order convex* provided that whenever $x \leqslant y \leqslant z$ for some x and z in K, the point y also lies in K. The order convex sets form the *order alignment* on X. The class of all alignments which arise as an order alignment for *some* partial order of the underlying set forms a variety denoted $\langle PO \rangle$. Notice that order alignments do *not*, in general, determine the ordering of the underlying space--even up to reversal.

The crux of the assertion that $\langle PO \rangle$ is a variety is this: if every finite subspace of an aligned space admits a local ordering which determines the alignment relative to that finite subspace, then the whole space admits a global order which determines the whole alignment. The demonstration of course uses the finitary property in an essential way and provides one illustration of the many ways the finitary property can be used to piece local information together to get global information.

A subset D of an ordered set X is a *downset* (or "order ideal") if for each x in D, all points y ⩽ x are also in D. These form the *downset alignment* on X. We have noted one occurrence of a downset alignment already-- namely, varieties correspond to downsets in the ordered space FinALN.

If the order on X is total (i.e., any two elements in X are comparable), then the above alignments are of enough importance to merit special attention. An order alignment induced by a total order is a *total order alignment*; a downset alignment of a total order is a *monotone* alignment. These two types of alignments are varieties denoted ⟨TO⟩ and ⟨Mon⟩, respectively.

B. Matroids and Antimatroids

Undoubtedly, the matroids (or "combinatorial geometries") form the best known and most thoroughly studied abstract class of alignments [4, 7, 58, 63, 64]. Although matroids are often required to be finite, that restriction is unnecessary (even undesirable) here. Among the numerous equivalent ways of defining matroids [63, 64], the definition in terms of hull operators seems most appropriate here. A *matroid* is an aligned space $(X, \mathcal{L})$ satisfying the *exchange law*: for any $L \in \mathcal{L}$ and $x, y \notin L$,

$$x \in \mathcal{L}(L \cup y) \quad \text{implies} \quad y \in \mathcal{L}(L \cup x)$$

The affine alignment on any vector space is a matroid, and there are numerous other examples [4, 63]. The class of matroids is a variety of alignments.

An *antimatroid* [33, 38] is an aligned space satisfying the *anti-exchange law*: for any $L \in \mathcal{L}$ and $x \neq y$ not in L,

$$x \in \mathcal{L}(L \cup y) \quad \text{implies} \quad y \notin \mathcal{L}(L \cup x).$$

As with the exchange law, this condition has numerous equivalents [8, 20, 26, 28, 38]. One of them is the finite Krein-Milman property: every polytope is the convex hull of its extremepoints (see Section 2D). Alignments satisfying the anti-exchange law are said to be *extremely detachable* (ED) [28], and the class of all such alignments forms a variety denoted ⟨ED⟩.

The variety ⟨ED⟩ of antimatroids includes

1. the ordinary alignments on real vector spaces

2. all order alignments

3. all downset alignments

4. chordal graphs with the monophonic alignment (recall (2.2.ii); see [37])

5. ptolemaic graphs [27] with the geodesic alignment [37]

6. hereditarily unicoherent locally connected continua with the alignment of all connected subsets

7. semilattices with the semilattice alignment

A *(join) semilattice* [2, 8, 31] is an ordered set in which each pair of elements has a supremum. The *semilattice alignment* consists of all subsemilattices--subsets closed under formation of pairwise suprema--of the underlying semilattice.

C. Separation Axioms

Within alignment theory, the basic separation axioms roughly parallel those of topology, using instead of open sets the notion of a *hemispace (biconvex set)*: a convex set with convex complement. Two sets are *separated* by a hemispace if one is contained in the hemispace and the other is contained in its complement. The basic separation axioms [23, 29, 32] are these:

S_0: distinct points have distinct hulls

S_1: points are convex

S_2: distinct points can be separated by a hemispace

S_3: any convex set can be separated from any point not in it by a hemispace

S_4: any two disjoint convex sets can be separated by a hemispace.

Axioms S_0, S_1, S_2, and S_3 define varieties. Axiom S_4 does not. Although it is finitary, it fails (like topological normality) to be inherited by subspaces. (Figure 1 shows a subspace of the plane that is *not* S_4.)

The following spaces satisfy *all* of the above separation axioms:

1. the ordinary alignment on real vector spaces

2. semilattices with the semilattice alignment

FIGURE 1

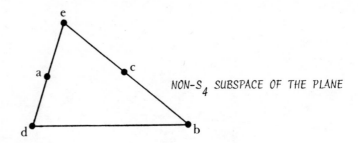

NON-S_4 SUBSPACE OF THE PLANE

3. trees and block graphs [24] with the geodesic alignment [37]

4. distributive lattices with the *order lattice* alignment of all order convex sublattices [2]

5. total order alignments

6. the affine alignment over GF(2), the two-element field

Partial order alignments do not, in general, satisfy S_4, although they always satisfy the other axioms. (In the ordered set whose Hasse diagram is shown in Figure 2, the order convex sets {a, b} and {c, d} cannot be separated.) All coaffine alignments satisfy S_0 through S_3.

D. Dimension Parameters

By a *dimension parameter* we shall mean a mapping δ which associates to each aligned space a nonnegative integer or ∞ such that

FIGURE 2

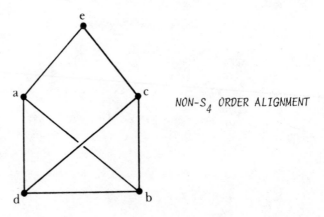

NON-S_4 ORDER ALIGNMENT

DP1. δ assigns the same value to isomorphic spaces

DP2. if $(Y,\mathcal{L}_2)$ is a subspace of $(X,\mathcal{L}_1)$, then $\delta(Y,\mathcal{L}_2) \leqslant \delta(X,\mathcal{L}_1)$

DP3. for any space $(X,\mathcal{L}_1)$, $\delta(X,\mathcal{L}_1) = \sup\{\delta(E,\mathcal{L}_1|E): $ E a finite subset of X}

It is clear from the definition that a dimension parameter δ is equivalent to a chain of varieties $V_n(\delta) = \langle \delta \leqslant n \rangle$ given by

$$\langle \delta \leqslant n \rangle = \{(X,\mathcal{L}): \delta(X,\mathcal{L}) \leqslant n\}$$

Examples of dimension parameters include Carathéodory number, rank, valence, height (all defined below), as well as cardinality $|X|$.

In an aligned space $(X,\mathcal{L})$, a set S is *irredundant over* a point p if $p \in \mathcal{L}(S)$ but $p \notin \mathcal{L}(T)$ for each proper subset T of S. By the finitary property irredundant sets are finite. The *Carathéodory number* c(X) is the supremum of the cardinalities of the irredundant subsets of X. As easily checked, c is a dimension parameter. The Helly and Radon numbers, while important invariants, are not dimension parameters: when passing to subspaces, one may delete required intersection points and thus increase these invariants. In semilattices, the Carathéodory number of the semilattice alignment is a well-established invariant known as the *breadth* (see p. 99 of [2] and p. 38 of [8]).

A subset I of an aligned space is *independent* if for each $x \in I$, $x \notin \mathcal{L}(I \setminus x)$. The *rank* r(X) is defined as the supremum of the cardinalities of the independent sets. Since any irredundant set is necessarily independent, $c(X) \leqslant r(X)$ for any X. The following result lists several useful equivalent formulations of the rank. Note that property (iii) says that the rank of an aligned space is the same as the Carathéodory number (breadth) of the (meet) semilattice of convex sets.

(4.1) For any aligned space $(X,\mathcal{L})$, the following are equivalent:

   (i)   rank (X) $\leqslant$ n

   (ii) for any finite set E, there is $F \subseteq E$ with $|F| \leqslant n$ and $\mathcal{L}(E) = \mathcal{L}(F)$

   (iii) for any finite family (L_i) $i \in I$ of convex sets, there is a subfamily (L_i) $i \in J$ of at most n of these with $\cap\{L_i: i \in I\} = \cap\{L_i: i \in J\}$

(iv) given any (n + 1)-copoints in X, one of them contains the intersection of the others

By a *copoint attached* at a point p of an aligned space X, we mean any
maximal convex subset of X \ p. By alignment axiom A3, copoints exist and
every convex set is an intersection of the copoints containing it. Thus copoints are of obvious significance in alignments, and as illustrated by condition (iv) of the theorem above, many properties can be formulated in terms
of copoints.

In the literature, copoints are known in various contexts as "completely
(meet)-irreducibles" [8, p. 43], "absolutely irreducible ideals" [51], "semi-
spaces" [21], and "hypercones" [42]. The term "copoint," which is borrowed
from matroid theory [63, p. 48], will be used here not only because it is
short and geometrically appropriate in general, but also because there is a
general "duality" theory in which copoints become "dual" points [30].

The *valence* $\mathrm{val}_p(X)$ *at a point* p in a space X is the number of copoints
at p. For trees in the geodesic alignment, this valence coincides with graph
theoretic degree. In finite semilattices, valence at a nonmaximal point is
the number of upper covers of that point [33]. The *(local) valence* val(X)
of X is the supremum over all $p \in X$ of $\mathrm{val}_p(X)$. This defines a dimension
parameter. Indeed, it is just one of many possible dimension parameters
definable using the *copoint intersection properties* CIP(n,k) introduced in
[36]. Since val(X) $\leqslant$ n is equivalent to CIP(n + 1, 1), it follows from Prop-
ersition 6 of [36] that c(X) $\leqslant$ val(X) for any X.

The *height* ht(X) of an aligned space is the maximum number of proper
convex sets in a chain. This defines a dimension parameter with r(X) $\leqslant$
ht(X).

E. Free Alignments and Quasicircuits

There are several special series of aligned spaces still requiring definition. The *free alignment* on any set X is simply the power set Pow(X)--
i.e., every set is convex. The free aligned space on n points will be denoted by F(n). For each integer k $\geqslant$ 0, the k-*free alignment* on a set X consists of all subsets of k or fewer points of X together with X itself. The
k-free alignment on n points will be denoted $F_k(n)$. Evidently, for k $\geqslant$ n - 1,
$F_k(n) = F(n)$.

For integers $n \geqslant k \geqslant 0$, the *quasicircuit* $Q_k(n)$ is defined to be the a-lignment on an n point set in which each subset of n - 2 or fewer points is convex and exactly k subsets of n - 1 points are convex [32]. Evidently, $Q_n(n) = F(n)$. $Q_{n-1}(n)$ may be visualized as the vertices of an (n - 2)-simplex together with an interior point. $Q_0(n) = F_{n-2}(n)$ is a true circuit in the matroid sense: it is the unique matroid of rank n - 1 on n points.

For each k, the k-free alignments form a variety ⟨k-free⟩. The class of all free alignments forms a variety ⟨free⟩. The quasicircuits together with the free alignments form a variety, but the main interest in the quasi-circuits is the role they play in characterizations by forbidden subspaces (cf. Section 7).

5. CONSTRUCTIONS WITH ALIGNMENTS AND VARIETIES

The list of examples of alignments can be further augmented by means of several natural constructions. In terms of varieties, we shall be interested in knowing what varieties are closed under which of the constructions. Table I summarizes some information along these lines. In Section 3 the construction of subspaces was introduced, and, by definition, every variety is closed under this construction.

TABLE I. Closure Properties of Varieties

Variety	Product-closed	Sum-closed	Join-closed	Contrac-tible	Lattice-closed (Section 6C)	Principal
⟨PO⟩	no	yes	no	no	no	yes
⟨TO⟩	no	no	no	no	no	yes
⟨monotone⟩	no	no	no	yes	no	yes
⟨matroids⟩	no	yes	no	yes	no	yes
⟨ED⟩	no	yes	yes	yes	no	yes
S_0, S_1, S_2, S_3	yes	yes	yes	no	no	yes
⟨Carath ⩽ n⟩	no	yes	no	yes	no	yes
⟨rank ⩽ n⟩	no	no	no	yes	yes	yes
⟨k-free⟩	only for k = 0	only for k = ∞	all k	all k	no	yes

A. Products, Sums, and Joins

   If $(X, \mathcal{L}_1)$ and $(Y, \mathcal{L}_2)$ are aligned spaces, then

$$\mathcal{L}_1 \otimes \mathcal{L}_2 = \{L \times M: \ L \in \mathcal{L}_1 \text{ and } M \in \mathcal{L}_2\}$$

is an alignment on $X \times Y$ (cf. [15, 29, 36, 48, 51, 52]). The space $(X \times Y,$ $\mathcal{L}_1 \otimes \mathcal{L}_2)$ is the *product* of $(X, \mathcal{L}_1)$ and $(Y, \mathcal{L}_2)$. Products of any number of factors can be formed similarly by first forming the collection of all "boxes" --cartesian products of one convex set from each factor--and then taking the alignment they generate. (In general, the boxes themselves form an align- ment only if the number of factors is finite [29].)

   Note that a product of d copies of the real line R does *not* yield the ordinary alignment on R^d, but rather the *box alignment* on R^d in which the only "convex" sets are boxes with sides parallel to the coordinate axes. This alignment has been studied extensively by Eckhoff [15] and others [36, 48, 52].

   Other structures fare better under products. The geodesic alignment on a cartesian product of graphs is the product of the geodesic alignments on the factors [37]. The order lattice alignment on a product of lattices is the product of the order lattice alignments on the factors. Since the or- dinary alignment on R *is* an order lattice alignment (for the usual order), the box alignment on R^d should also be an order lattice alignment. Indeed it is, with R^d made into a lattice by coordinatewise ordering.

   Using the finitary property, one can show that a variety is closed un- der the formation of *arbitrary* products provided it contains the product of each *pair* of its *finite* spaces. Such a variety is said to be *product-closed*. The varieties $\langle S_0 \rangle$, $\langle S_1 \rangle$, $\langle S_2 \rangle$, and $\langle S_3 \rangle$ are product-closed. As we shall see (Section 4C), being product-closed is a strong requirement.

   A less demanding notion is that of sum. The *sum* of a collection of a- ligned spaces is obtained by representing them on disjoint underlying spaces and then aligning the disjoint union with all possible unions of one convex set from each summand. A variety is *sum-closed* if it is closed under the formation of arbitrary sums. Again it is sufficient, by the finitary prop- erty, that the sum of each *pair* of *finite* spaces in the variety again be in the variety. Being sum-closed is a weak condition, enjoyed by most of the varieties we have discussed.

If $(\mathcal{L}_\gamma)$, $\gamma \in \Gamma$, is a collection of alignments on a set X, then the small-est alignment $\mathcal{R}$ on X containing all $\mathcal{L}_\gamma$'s is the *join* of the $\mathcal{L}_\gamma$'s in the lat-tice of all alignments on X. It can be shown [29], that if $\mathcal{R}$ is the join of $(\mathcal{L}_\gamma)$, $\gamma \in \Gamma$, then

$$(5.1) \quad \mathcal{R}(E) = \bigcap_{\gamma \in \Gamma} \mathcal{L}_\gamma(E)$$

for all *finite* subsets E of X. If this formula actually holds for *all* sub-sets E of X, we shall say that $\mathcal{R}$ is a *strong join* of $(\mathcal{L}_\gamma)$, $\gamma \in \Gamma$. Further,

(5.2) The ordinary alignment on any real vector space is a strong join of total order alignments.

(5.3) The order alignment on any partially ordered set is a strong join of total order alignments.

Fact (5.2) is essentially a version of the separation properties of ordinary convex sets. (5.3) is a consequence of the fact that any partial order, as a set of ordered pairs, is the intersection of the total orders extending it. The occurrence of strong joins is not a coincidence, but an illustration of a general principle involving varieties.

(5.4) If an alignment is a join of alignments from a variety V, then it is actually a strong join of alignments from V.

While it can be shown that the join of any *finite* number of alignments is strong [29, p. 20], this need not be the case for infinite joins. Indeed, the free alignment on an infinite set X is the join, but *not* the strong join, of the k-free alignments (k = 1, 2, 3, ...) on X.

A variety is *join-closed* if it is closed under the formation of arbi-trary joins. Again, for a variety to be join-closed it suffices for it to contain pairwise joins on finite spaces.

B. The Category of Aligned Spaces and Minors

The class of all aligned spaces can be made into a category *ALN* in which the morphisms from $(X, \mathcal{L}_1)$ to $(Y, \mathcal{L}_2)$ are maps from X to Y such that the pre-

image of each convex set in Y is convex in X. The notions of product and
sum introduced above are the correct ones for this category. The category
ALN also possesses inductive and projective limits.

(5.5) Any variety is closed under arbitrary inductive and projective limits.

Since by the finitary property, every aligned space is the inductive limit
of its finite subspaces, closure under inductive limits is essentially an
equivalent formulation of variety axiom V3.

The category ALN also contains quotients. For the most part, these are
not well-enough behaved to play a major role in the structure theory of alignments. There is, however, one type of quotient that is of general interest.
If K is a convex set in an aligned space $(X,\mathcal{L})$, then the *contraction of* $(X,\mathcal{L})$
over K has $X \setminus K$ as underlying set and

$$\mathcal{L} \, / \, K = \{L \setminus K: \ L \in \mathcal{L} \text{ and } K \subseteq L\}$$

as alignment. A subspace of a contraction of X (or, equivalently, a contraction of a subspace) is a *minor* of X. (This term is borrowed from matrix
theory via matroids [4, 58, 63].) It is not difficult to check that a minor
of a minor of X is again a minor of X. A variety closed under the formation
of arbitrary contractions (and hence of minors) is *contractible*.

As with products, sums, and joins, there is also a finite sufficient
condition for contractibility.

(5.6) A variety V is contractible iff for each finite space $(E,\mathcal{L})$ in V and
each p in E, the contraction of E over $\mathcal{L}(p)$ is in V.

C. Product-Closed Varieties

The relative strength of product-closure is indicated by the results
below.

(5.7) Any product-closed variety is join-closed.

(5.8) A product-closed variety other than ⟨1 pt⟩ or ⟨0-free⟩ is sum-closed.

(5.9)   The only product-closed variety containing both $Q_o(2)$ and $Q_1(2)$ is $\langle$all$\rangle$, the variety of all aligned spaces.

(5.10) The only varieties closed under both products and contractions are $\langle$1 pt$\rangle$, $\langle$0-free$\rangle$, and $\langle$all$\rangle$.

As an illustration of some of the techniques involved, we shall actually prove these results.

   Proof of (5.7).   The join of two alignments $\mathcal{L}_1$ and $\mathcal{L}_2$ on a space X is isomorphic to the diagonal subspace of X × X with the product alignment $\mathcal{L}_1 \otimes \mathcal{L}_2$. □

   As an aid in (5.8) and (5.10), the following lemma is useful.

(5.11) If V is a product-closed variety other than $\langle$1 pt$\rangle$ or $\langle$0-free$\rangle$, then V contains $Q_2(2) = F(2)$.

   Proof.   Since any 0-free alignment is a subspace of products of $Q_o(2)$, V is *not* contained in $\langle$0-free$\rangle$ or it would be either $\langle$1 pt$\rangle$ or $\langle$0-free$\rangle$. Hence V contains a space X with a nonempty, proper convex set K.   Pick any $x \in K$ and any $y \in X \setminus K$.   Then $\{x\}$ is convex in the subspace $\{x,y\}$ of X, so this subspace is either $Q_1(2)$ or $Q_2(2)$.
   In the latter case we are done.   In the former, V contains $Q_1(2)$ which can be represented in two ways on $\{0,1\}$--one with 0 convex, the other with 1 convex.   The join of these representations is $Q_2(2)$, which by (5.7) must be in V as desired.   □

   Proof of (5.8).   Given spaces X and Y, pick and fix $p \in X$ and $q \in Y$. The sum of X and Y is then isomorphic to the following subspace of the product $X \times Q_2(2) \times Y$:

   $(X \times \{0\} \times \{q\}) \cup (\{p\} \times \{1\} \times Y)$ □

Theorem (5.9) is an immediate consequence of the following stronger result:

(5.9') Any aligned space $(X, \mathcal{L})$ is isomorphic to a subspace of a product of copies of $Q_0(2)$ and $Q_1(2)$.

Proof. Take $\Gamma = X \cup \mathcal{L}$ as index set and let $Z = \underset{\gamma \in \Gamma}{\times} \{0,1\}_\gamma$ where the factors indexed by points in X are taken isomorphic to $Q_0(2)$ and the factors indexed by sets in $\mathcal{L}$ are taken isomorphic to $Q_1(2)$ with $\{1\}$ convex. For each x in X, define ϕ_x in Z as follows:

for $y \in X \subseteq \Gamma$,
$$\phi_x(y) = 1 \quad \text{if } x = y$$
$$\phi_x(y) = 0 \quad \text{if } x \neq y$$

for $K \in \mathcal{L} \subseteq \Gamma$,
$$\phi_x(K) = 1 \quad \text{if } x \in K$$
$$\phi_x(K) = 0 \quad \text{if } x \notin K$$

The factors $Q_0(2)$ for each y in X ensure that the map $x \to \phi_x$ from X to Z is one-to-one. The factors $Q_1(2)$ for each K in $\mathcal{L}$ ensure that it is an isomorphism with a subspace of Z. $\square$

Proof of (5.10). Suppose V is a contractible, product-closed variety other than $\langle 1 \text{ pt} \rangle$ or $\langle 0\text{-free} \rangle$. By lemma (5.11), V contains F(2) as the free alignment on $\{0,1\}$. Now V contains the product $F(2) \times F(2)$. Contract this product over the convex point $(0,0)$. In the contraction, the hull of $(1,1)$ contains $(0,1)$, but not vice versa. Hence $Q_1(2)$ is a minor of $F(2) \times F(2)$ and thus is in V.

Now contract the convex edge $\{(0,0), (0,1)\}$ of $F(2) \times F(2)$. This space is $Q_0(2)$ which must then be in V. Since V contains $Q_0(2)$ and $Q_1(2)$, V = $\langle \text{all} \rangle$ by (5.9). $\square$

On the basis of these results one might suspect that there are really rather few product-closed varieties. On the contrary, it can be shown that the lattice of product-closed varieties contains a copy of a complete Boolean algebra on a (countably) infinite set. It would be of some interest to obtain a complete description of the lattice of product-closed varieties.

All known product-closed varieties are defined by some type of "separation" axiom. Is this possibly always the case for some reasonable notion of "separation axiom?"

D. Join-Closure of Varieties

For any two varieties V and W, let V · W be the class of all aligned spaces $(X,\mathcal{L})$ where $\mathcal{L}$ is the join of an alignment in V with an alignment in W. We set $V^2 = V \cdot V$ and define analogously V^n to be the class of all alignments obtainable as the join of n (or fewer) alignments from V. Likewise, V^∞ denotes those alignments which are joins of some (maybe infinite) collection of alignments in V. It can be shown that all of these constructions yield varieties. (Axioms V1 and V2 are trivially satisfied; the difficulty lies in establishing V3.)

Thus V^∞ is the *join-closure* of V (i.e., it is the smallest join-closed variety containing V). Of particular interest is the join-closure $\langle TO \rangle^\infty$. As already noted in (5.2) and (5.3) the ordinary alignments on real vector spaces as well as all (partial) order alignments belong to $\langle TO \rangle^\infty$. This suggests that a study of $\langle TO \rangle^\infty$ would be a reasonable step toward a characterization of the finite subspaces of Euclidean space. This approach has in fact been taken implicitly by Goodman and Pollack in their studies of the geometry of "allowable sequences." (See [18] and Pollack's chapter in these *Proceedings*.)

While $\langle TO \rangle^\infty$ consists of those spaces determined by "two-sided" total orders, $\langle Mon \rangle^\infty$ is the related class determined by "one-sided" total orders. Although no simple characterization of $\langle TO \rangle^\infty$ is known, the join-closure of the monotone alignments is the variety of antimatroids [33, 38]:

$$\langle Mon \rangle^\infty = \langle ED \rangle$$

Another simpler result along these lines is

$$\langle rank \leqslant 1 \rangle^\infty = \langle all \rangle$$

Indeed, any maximal chain in any alignment is a rank 1 subalignment, and every alignment is certainly the join of its chains.

Notice that with every variety V we have a chain of varieties $V^n \subseteq V^{n+1}$ whose supremum (in the lattice of varieties) is the join-closure V^∞. These chains of varieties provide abundant additional examples of dimension parameters:

$$j_V(X) = \min\{n: \ X \in V^n\}$$

6. RELATED VIEWS: LOGIC, LATTICES, AND ALGEBRAS

A. Syntax of Varieties

In Section 2 alignments were introduced in two ways: as systems of sets and as hull operators. Also, two ways of viewing varieties were noted: as classes of spaces and as downsets in the ordered set FinALN. Loosely speaking, these views are *semantic* in that they directly involve specific structures rather than the logical forms of the properties defining them. Here we will briefly take a *syntactical* view, giving attention to the logical structure of properties of alignments. The exposition is simply a working out of well-known principles in the first order logic of relational structures for the case of alignments. Details may be found in the standard references [5, 6, 19].

To know an alignment $\mathcal{L}$ on a set X, it is necessary and sufficient to know for each subset S of X and each point p in X whether or not the relation $p \in \mathcal{L}(S)$ holds. In view of the finitiary property (2.1), it suffices to know this when S is finite. Hence, we may associate with any aligned space $(X, \mathcal{L})$ a system $\Omega = (\omega_n(\mathcal{L}))_{n=1}$ of n + 1-ary relations (i.e., subsets of X^{n+1}) defined by

$$(z, x_1, \ldots, x_n) \in \omega_n(\mathcal{L}) \quad \text{iff} \quad z \in \mathcal{L}(z_1, \ldots, x_n) .$$

These relations uniquely determine the alignment and enjoy the following properties:

ARS1. $(z,z) \in \omega_1$ for all z

ARS2.   if $(z, x_1, \ldots, x_n) \in \omega_n$, then for any permutation π of 1, $\ldots$, n, $(z, x_{\pi(1)}, \ldots, x_{\pi(n)}) \in \omega_n$

ARS3.   if $(z, x_1, \ldots, x_n) \in \omega_n$ and $x_{n-1} = x_n$, then $(z, x_1, \ldots, x_{n-1}) \in \omega_{n-1}$

ARS4.   if $(z, x_1, \ldots, x_n) \in \omega_n$, then for any y, $(z, x_1, \ldots, x_n, y) \in \omega_{n+1}$

ARS5.   for any n × n matrix of elements (x_{ij}) and elements $y_1, \ldots, y_n$ and z, if $(y_i, x_{i1}, x_{i2}, \ldots, x_{in}) \in \omega_n$ for all j and $(z, y_1, \ldots, y_n) \in \omega_n$, then $(z, x_{11}, \ldots, x_{ij}, \ldots, x_{nn}) \in \omega_{n^2}$

Conversely, any sequence of relations $\Omega = (\omega_n)$ satisfying ARS1-ARS5 arises as $\omega_n(\mathcal{L})$ for some (unique) alignment $\mathcal{L}$--just interpret $(z, x_1, \ldots, x_n) \in \omega_n$ to mean "z is a convex combination of $x_1, \ldots, x_n$" and define the alignment of convex sets accordingly. Thus alignments may be identified with relational structures satisfying ARS1-ARS5.

For convenience, we shall henceforth drop the ω_n-notation and write simply $z \in \mathcal{L}(x_1, \ldots, x_n)$. The logical formulas with which we shall be concerned are those obtainable from atomic formulas of the kind $z \in \mathcal{L}(z_1, \ldots, x_n)$ and $x = y$ by using logical connectives $\wedge$ (and), $\vee$ (or), $\ulcorner$ (not), $\Rightarrow$ (implies).

For example, separation axiom S_0 can be stated as

$$S_0: \quad \forall x \, \forall y \quad x \in \mathcal{L}(y) \wedge y \in \mathcal{L}(x) \Rightarrow x = y$$

Likewise, spaces with Helly number $\leqslant n$ are those satisfying

$$H_n: \quad \forall x_0, x_1, \ldots, x_n \, \exists p \quad \bigwedge_{i=0}^{n} p \in \mathcal{L}(x_0, \ldots, \hat{x}_i, \ldots, x_n)$$

where $\hat{\ }$ denotes "omit." (H_n actually says any $n + 1$ points have nonempty core [30, 36, 53].)

Both of these statements are *first order*--that is, quantified over elements only and not over sets. The first defines a variety and the second does not. This is a consequence of a general result of Robinson and Łos (see Theorem VI.2.8 of [6]) which for alignments states

(6.1) Varieties of alignments are precisely the *universal classes* A, that is, classes defined by universally quantified first order axioms.

This may not always appear to be the case. For example, the exchange law (Section 4B) appears to be quantified over sets and separation axiom S_3 (Section 4C) appears to postulate the *existence* of hemispaces. But universally quantified first order equivalents can be found without much trouble. The reader may wish to try his hand at this. As a rule of thumb, assertions about *all* convex sets may be reduced via the finitary property to assertions about polytopes. The "general" polytope can then be treated as the hull of a finite number of variables. The result may be a sequence of axioms rather than just one. As an illustration, here are axioms for the variety ⟨Carath $\leqslant n$⟩--we take all axioms C_k with $k > n$:

$$C_k: \quad \forall x_1, \ldots, x_k \; \forall p \qquad p \in \mathcal{L}(x_1, \ldots, x_k)$$

$$\Rightarrow \bigvee_{i=1}^{k} p \in \mathcal{L}(x_1, \ldots, \hat{x}_i, \ldots, x_k)$$

Using the fact that varieties can be defined in terms of minimal forbidden subspaces (Section 3), it is easy to see why (6.1) holds. Indeed, suppose $(E, \mathcal{L})$ is a finite aligned space. The requirement that a space contain E as a subspace can be axiomatized by taking one variable x_i for each point of E and postulating

$$(E \subseteq X): \quad \exists x_1, \ldots, x_n \quad (\bigwedge_{i \neq j} x_i \neq x_j) \wedge \phi(x_1, \ldots, x_n)$$

where ϕ is a long conjunction of all relations of either the form $z \in \mathcal{L}(F)$ or the form $z \notin \mathcal{L}(F)$--according to which holds in E. The negation of $(E \subseteq X)$ is universally quantified and states that E is *not* a subspace of X. Thus any variety can be defined by a list of axioms, one for each minimal forbidden subspace. Since any finite number of axioms can always be combined by conjunction into a single axiom, we obtain

(6.2) A variety has only finitely many minimal forbidded subspaces iff it can be defined by a *single* universal first order axiom.

A further illustration of the important correspondence displayed in (6.1) and (6.2) between semantics and syntactics is the well-known fact (see Theorem VI.4.3 of [6]) that product-closed varieties can be defined by Horn sentences. It would be interesting to have similar syntactical descriptions of join-closed varieties and of contractible varieties.

B. Universal Algebras

We have just seen how alignments can be viewed as a type of relational system. They are also closely related to another, widely studied type of relational structure: universal algebras [6, 19]. (A concise introduction can be found in [16].) Universal algebras are those relational systems in which all the relations are functions, that is, *operations* on the underlying set. Rings, groups, lattices, semilattices, and vector spaces over a fixed field are examples.

For any universal algebra, the family of subalgebras (subsets closed under all the operations) forms an alignment, and it is wellknown that any alignment can be represented as the subalgebras of *some* algebra structure on the underlying set (hence the traditional description of alignments as "algebraic closure systems"). However, there is no canonical way to assign an algebra structure to a general alignment, so it is difficult to carry over information from the theory of universal algebras into the general theory of alignments. Nonetheless, many important specific classes of alignments (affine, semilattice, order lattice) arise from important classes of universal algebras, and this author believes that more use could and should be made in "axiomatic convexity" of examples from universal algebra.

There is also a well-developed and highly successful theory of varieties of algebras. (See [16] on this point.) Syntactically, a variety of algebras is an *equational class*, defined by axioms that are simple equations of polynomials (compositions) in the basic operations. Semantically, varieties of algebras are classes closed under the formation of subalgebras, products, and quotients.

Although it would be nice to emulate the theory of varieties of algebras, one cannot use for alignments the same semantic requirements. We have already seen in (5.10) that there are only *three* classes of alignments closed under subspaces, products, and contractions, hardly a discriminating classification scheme. A major difference lies in the fact that *subspaces* are far more general than *subalgebras*--any subset can be a subspace, but subalgebras correspond to *convex* subsets.

For a discussion of ordinary convexity from the universal algebra viewpoint, see [47].

C. Algebraic Lattices and Lattice-Closed Varieties

Every alignment as a family of sets is a lattice under set inclusion in which infimum is intersection and supremum is convex hull of the union. It has long been known exactly which lattices arise as alginments. (See Theorems 8 and 8' on page 187 of [2].) They are those lattices which are *complete* (arbitrary sups and infs exist) and *algebraic*, defined as follows. An element c of a lattice L is *compact* iff whenever $c \leqslant \sup S$ for some $S \subseteq L$, there is a finite $F \subseteq S$ with $c \leqslant \sup F$. A lattice is *algebraic* if each element is a sup of compact elements. Note that the compact elements of an alignment are precisely the polytopes.

TABLE 2. Lattice Properties of Varieties

all alignments	complete algebraic
S_1: points convex	atomic
matroids	upper semimodular
antimatroids ⟨ED⟩	dually locally distributive (see Chapter 7 of [8])
⟨Carath ≤ 1⟩	distributive
⟨rank ≤ n⟩	breadth ≤ n

It has long been popular to classify and identify various types of alignments by properties of their lattices. (See especially [40].) Table 2 contains several classes of alignments and their characteristic lattice properties. (Note that ⟨Carath ≤ 1⟩ is precisely the class of topologies which
are also alignments.) When points are convex, this classification of alignments by their lattices is quite satisfactory since S_1 spaces are determined
up to isomorphism by their lattices of convex sets.

When points are *not* convex, the correspondences become ambiguous. Non-
isomorphic--and in fact geometrically very different--aligned spaces can have
isomorphic lattices of convex sets. As shown by (6.6) below, every *finite*
aligned space can be realized as a subspace of an alignment lattice-isomorphic to a free alignment. One could eliminate the ambiguity--as is often
done (e.g., [65])--by requiring points to be convex. But this also elimi-
nates some pleasing structural considerations, such as the characterization
of antimatroids as joins of monotone alignments (Section 5D) and the simple
forbidden minor characterization of matroids and antimatroids (Section 7).

It seems best to recognize the validity and usefulness of both classification methods for alignments: by lattices and by varieties. There is,
in fact, an interesting area of overlap. A variety V is *lattice-closed* if
for each space $(X, \mathcal{L}_1)$ in V, every space $(Y, \mathcal{L}_2)$ with $\mathcal{L}_2$ lattice-isomorphic
to $\mathcal{L}_1$ is also in V. Since the variety ⟨ht(X) ≤ n⟩ was defined in terms of
the lattice of convex sets, it is certainly lattice-closed. By (4.1.iii),
the variety ⟨rank ≤ n⟩ can be defined by a lattice property, so it is also
lattice-closed. Further examples arise from the following simple simple observation.

(6.3) If V and W are lattice-closed, then so is V · W.

The next results indicate some restrictions on lattice-closed varieties.

(6.4) There are only two varieties that are both lattice-closed and join-closed, namely, ⟨0-free⟩ and ⟨all⟩.

(6.5) If V ≠ ⟨all⟩ is a lattice-closed variety, then there is an N > 0 such that rank(X) ⩽ N for all X in V.

This second result follows at once from (6.6) below and the observation that a space of rank ⩾ n contains the free alignment F(n) as a subspace. (The relative alignment on any independent set is free.)

(6.6) Suppose each point of the aligned space $(X, \mathcal{L}_1)$ lies in all but finitely many $\mathcal{L}_1$-convex sets. Then $(X, \mathcal{L}_1)$ can be embedded as a subspace of a space $(Y, \mathcal{L}_2)$ where $\mathcal{L}_2$ is lattice-isomorphic to a free alignment. That is, as a lattice, $\mathcal{L}_2$ is a complete Boolean algebra.

Proof of (6.6). Take $Y = X \cup \mathcal{L}_1$. For each point x in X, let

$$\Phi(x) = \{L \in \mathcal{L}_1 : \quad x \notin L\}$$

Align Y with families M ⊆ Y such that

$$x \in M \quad \text{iff} \quad \Phi(x) \subseteq M$$

for each x in X. Since $\Phi(x)$ is finite for each x, the collection $\mathcal{L}_2$ of all such M is an alignment on Y. Since for each subfamily $\mathcal{R} \subseteq \mathcal{L}_1$,

$$\mathcal{L}_2(\mathcal{R}) = \mathcal{R} \cup \{x : \quad \Phi(x) \subseteq \mathcal{R}\}$$

one can show that $\mathcal{L}_2$ is isomorphic as a lattice to $\text{Pow}(\mathcal{L}_1)$. It is also not hard to see that $\mathcal{L}_1 = \mathcal{L}_2 \mid X$, as desired. □

The hypothesis of (6.6) admits all *finite* aligned spaces, but it also admits some infinite ones, for example, the monotone alignment on the positive in-

tegers. It would be interesting to know all spaces for which the conclusion
of (6.6) holds.

7. CHARACTERIZING VARIETIES BY FORBIDDEN SUBSTRUCTURES

As noted in Section 3, any variety is completely determined by its minimal forbidden spaces (MFS). For example, a space satisfies separation axiom
S_1 iff it contains neither of the quasicircuits $Q_0(2)$ and $Q_1(2)$. Also, a
space has rank $\leqslant$ n provided it has no subspace isomorphic to $F(n + 1)$.

(7.1) The MFS for the variety of matroids are the quasicircuits $Q_k(n)$ for
$0 < k < n$ and n arbitrary.

(7.2) There are seven MFS for $\langle$TO$\rangle$: namely, $Q_0(2)$, $Q_1(2)$, $Q_0(3)$, $Q_1(3)$,
$Q_3(3)$, $F(2) \times F(2)$, and the "Arrow" space [32].

The "Arrow" space is the subspace of four white nodes of the chordal graph
in Fig. 3 under the monophonic alignment.

Certainly the success of and interest in a characterization by forbidden
subspaces depends on the number and complexity of the MFS. Of the four examples cited above, three have only finitely many MFS. For such varieties,
the MFS characterization may be regarded as most effective. We shall say
that a variety is *finitely based* if it has only a finite number of MFS. Recall from (6.2) that this is equivalent to being definable by a single first

FIGURE 3

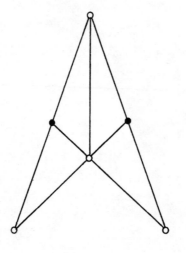

THE ARROW SPACE

order axiom. The following is a list of some of the known finitely based varieties:

$\langle S_0 \rangle$

$\langle S_1 \rangle$

$\langle monotone \rangle^k$ for all k

$\langle rank \leqslant n \rangle$ for all n

$\langle rank \leqslant 1 \rangle^k$ for all k

$\langle TO \rangle$

$\langle PO \rangle \cap \langle rank \leqslant n \rangle$ for all n

$\langle S_3 \rangle \cap \langle rank \leqslant n \rangle$ for all n

$\langle matroid \rangle \cap \langle rank \leqslant n \rangle$ for all n

$\langle ED \rangle \cap \langle rank \leqslant n \rangle$ for all n

From the last four examples, one might be tempted to suspect that, for each n, every variety contained in $\langle rank \leqslant n \rangle$ is finitely based. This is indeed true for n = 1 although the proof is not as trivial as one might expect. For $n \geqslant 3$, the conjecture is false, however. As Allan Day has observed [10], the projective planes over finite fields with different characteristics form an antichain of rank 3 alignments in FinALN. Hence the variety of all alignments of rank $\leqslant 3$ which contain none of these planes as subspaces is not finitely based. The problem remains open for rank $\leqslant 2$.

Even if a variety is not finitely based, its list of MFS may not be all that bad. For example, as seen in (7.1) the MFS for matroids are easily described. For order alignments (see [32]), they are more complicated but still within reason. For $\langle Carath \leqslant n \rangle$, the MFS are awful: one essentially needs to describe every space with Carathéodory number n − 1 in order to describe the MFS of $\langle Carath \leqslant n \rangle$. For free alignments, the MFS are the quasi-circuits $Q_k(n)$ with k < n: not so bad, but certainly not as simple as the free alignments themselves.

As a means to measure this complexity of a variety, this author would like to propose the following construction. For any variety V, let $\langle MFS(V) \rangle$ be the variety generated by the minimal forbidden subspaces of V. Now let the *core* of V be the subvariety generated by the infinite spaces in $\langle MFS(V) \rangle$. From the Upward Löwenheim-Skolem Theorem (page 67 of [5]), one can show that

(7.3) The finite spaces in core(V) are precisely those which lie in infinitely many MFS of V.

From this, it follows that

(7.4) For any variety V, core(V) $\subseteq$ V. Also core(V) $\neq \phi$ iff V is not finitely based.

Thus the size of core(V) in V measures the effectiveness of MFS characterization for V. From (7.1), we have core⟨matroids⟩ = ⟨free⟩, only a very small portion of matroids. On the other hand, core⟨free⟩ = ⟨free⟩, indicating the ineffectiveness of MFS classification in this case. For ⟨PO⟩, the core contains only spaces arising from orders of height 3 or less, again a small portion of ⟨PO⟩.

In contractible varieties, one can simplify the characterizations by considering *minimal excluded minors* (recall Section 5B) instead of forbidden subspaces. In the variety of matroids, this has become the accepted method of characterization. (See Chapter 6 of [58] for a discussion and some nontrivial examples.) The examples below can be derived at once from the definitions.

(7.5) An aligned space is a matroid iff it has no minor isomorphic to $Q_1(2)$.

(7.6) An aligned space is an antimatroid iff it has no minor isomorphic to $Q_0(2)$.

(7.7) An aligned space is free iff it has no minor isomorphic to either $Q_0(2)$ or $Q_1(2)$.

(7.8) The minimal excluded minors for ⟨Carath $\leq$ n⟩ all have n + 2 points—hence there are only finitely many such excluded minors.

8. SOME OPEN PROBLEMS

Perhaps the most intriguing problem in the theory of varieties is the characterization of those finite aligned spaces which can occur as subspaces of Euclidean space with ordinary convexity. In the language of varieties, the question asked is this:

PROBLEM 1. Characterize the variety $\langle R^d \rangle$ generated by ordinary convexity on R^d.

As noted in Section 5D, the varieties $\langle R^d \rangle$ all lie in $\langle TO \rangle^\infty$. (Refer also to R. Pollack's chapter in these *Proceedings*.)

PROBLEM 2. Find an infinite class of MFS for $\langle R^d \rangle$ which lie in $\langle TO \rangle^\infty \cap \langle Carath \leq d + 1 \rangle$.

Additional problems are noted below.

PROBLEM 3. Describe the lattice of produce-closed varieties. (Section 5C)

PROBLEM 4. Do all product-closed varieties arise from "separation axioms" of some kind? (Sections 4C, 5C, and 6A)

PROBLEM 5. Under the operation $V \cdot W$ the class of varieties forms a commutative semigroup. What is the structure of this semigroup? (Section 5D)

PROBLEM 6. Give syntactic descriptions of contractible and join-closed varieties. (Sections 5B, 5D, and 6A)

PROBLEM 7. Characterize the variety $\langle TO \rangle^\infty$. (Section 5D).

PROBLEM 8. Is the variety $\langle TO \rangle^n$ finitely based for each n? (Sections 5D and 7)

PROBLEM 9. More generally, if V and W are finitely based, is $V \cdot W$ finitely based? That is, in reference to Problem 5, do the finitely based varieties form a subsemigroup? (Sections 5D and 7)

PROBLEM 10. Are all rank 2 varieties finitely based?

PROBLEM 11. Are all lattice-closed varieties finitely based?

PROBLEM 12. What is the structure of FinALN as an ordered set? (Section 3)

ADDENDUM

Jeff Kahn and J. P. S. Kung (*AMS Bull.* 3(1980), 857) have announced a classification of all varieties of matroids. Their notion of variety assumes contractible and several other properties not discussed here. A. J. Hoffman has kindly reminded me of Richard Rado's classic paper (*J. London Math. Soc.* 22(1947), 219-226) in which the connection between Helly's theorem and the Chinese Remainder Theorem is noted and investigated (cf. Section 2.A). Paul H. Edelman (*Alg. Universalis* 10(1980), 290-299) has independently shown that the variety of antimatroids is closed under joins (cf. Table I, Section 5) and that they are characterized as dually locally distributive lattices (cf. Table II, Section 6). A recent book by G. Gierz *et al.* (*A Compendium of Continuous Lattices*, Springer Verlag, Berlin, 1980) has considerable impact on the theory of alignments since alignments are a special class of continuous lattices. In fact, as J. D. Lawson has noted (private communication), continuous lattices may be viewed as "fuzzy" alignments.

ACKNOWLEDGMENTS

The author is grateful to numerous people for their assistance and kind encouragement with this report. Space does not permit thanking them all individually here, but special acknowledgments are surely due to Rudolf Wille and the faculty and guests of Arbeitsgruppe I in Darmstadt for many valuable conversations, especially during the author's visit in 1979; to Stefan Foldes for pointing out the syntactic approach to varieties; to Hal Kierstead for assisting with an understanding of first order logic; and to the conference organizers, David Kay and Marilyn Breen, and to wife Dori, for their patience and encouragement during the writing of this report. The author is also grateful to the Alexander von Humboldt Stiftung for the opportunity of spending 1976-1977 at the University of Erlangen where this work was begun.

This report was prepared with partial support from NSF Grant MCS-8002543.

REFERENCES

1. C. Berge, *Graphs and Hypergraphs*, North-Holland Publ., Amsterdam, 1973.

2. G. Birkhoff, *Lattice Theory*, 3rd edition, Vol. XXV, AMS Colloquium Publ., Providence, R. I., 1967.

3. G. Birkhoff, and Frink, O., Representations of lattices by sets, *Trans. AMS* 64(1948), 229-315.

4. T. Brylawski, and Kelly, D. G., Matroids and combinatorials geometries, *Studies in Combinatorics, MAA Studies* Vol. 17, G. C. Rota, ed., (1978), 179-217.

5. C. C. Chang, and Keisler, H. J., *Model Theory*, Studies in Logic, Vol. 73, North-Holland Publ., Amsterdam, 1973.

6. P. M. Cohn, *Universal Algebra*, Harper and Row, New York, 1965.

7. H. H. Crapo, and Rota, G. C., *Combinatorial Geometries*, MIT Press, Cambridge, Mass., 1970.

8. P. Crawley, and Dilworth, R. P., *Algebraic Theory of Lattices*, Prentice-Hall, London, 1973.

9. L. Danzer, Grünbaum, B. and Klee, V. L., Helly's theorem and its relatives, *Proc. of the Symp. on Pure Math*, AMS Vol. 7 (Convexity) (1963), 101-180.

10. A. Day, Private Communication.

11. H. J. P. De Smet, On the convex closure operators in a field, compatible with the algebraic structure, *Bull. Soc. Math. Belgique* 26(1974), 261-273.

12. H. J. P. De Smet, Determination of the extremal points of the convex semigroup S_0 in the field F, *Bull. Soc. Math. Belgique* 26(1974), 363-370.

13. G. A. Dirac, On rigid circuit graphs, *Abh. Math. Seminar Univ. Hamburg* 25(1961), 72.

14. J. P. Doignon, Convexity in crystallographic lattices, *J. Geom.* 3(1973), 71-85.

15. J. Eckhoff, Der Satz von Radon in konvexen Produktstrukturen II, *Monatsh. Math.* 73(1969), 7-30.

16. T. Evans, Universal algebra and Euler's officer problem, *Am. Math. Monthly* 86(1979), 466-473.

17. D. R. Fulkerson, and Gross, D. A., Incidence matrices and interval graphs, *Pac. J. Math.* 15(1965), 835-855.

18. J. E. Goodman, and Pollack, R., On the combinatorial classification of nondegenerate configurations in the plane, *J. Comb. Th.* (a) 29(1980), 220 - 235.

19. G. Graetzer, *Universal Algebra*, Van Nostrand, Princeton, 1968.

20. R. L. Graham, Simonovits, M, and Sos, V. T., A note on the intersection properties of subsets of the integers, *J. Comb. Th.* (A) 28(1980), 107-110.

21. P. Hammer, Maximal convex sets, *Duke Math. J.* 22(1955), 103-106.

22. P. C. Hammer, Extended topology: Domain finiteness, *Indag. Math.* 25
     (1963), 200-212.

23. R. Hammer, Beziehungen zwischen den Saetzen von Radon, Helly and Cara-
     théodory bei axiomatischen Konvexitaeten, *Abh. Math. Sem. Univ. Hamburg*
     46(1977), 3-24.

24. F. Harary, *Graph Theory*, Addison-Wesley Publ., London, 1969.

25. S. P. R. Hebbare, A class of distance convex simple graphs, *Ars Com-
     binatorica* 7(1979), 19-26.

26. A. J. Hoffman, Binding constraints and Helly numbers, *Annals of NY Acad.
     Sci.*, Second Int'l. Conf. on Comb. Math. 319(1979), 284-288.

27. E. Howorka, A characterization of Ptolemaic graphs; survey of results,
     Proc. 8th SE Conf. Comb., *Graph Theory and Computing*, 355-361.

28. R. E. Jamison, A development of axiomatic convexity, *Tech. Report Nr.
     48*, Clemson Univ., 1970.

29. R. E. Jamison, *A General Theory of Convexity*, Dissertation, Univ. of
     Washington, 1974.

30. R. E. Jamison, A general duality between the theorems of Carathéodory
     and Helly, *Technical Report #205*, Clemson University, 1975.

31. R. E. Jamison, Tietze's Convexity Theorem for semilattices and lattices,
     Semigroup Forum 15(1978), 357-373.

32. R. E. Jamison, A convexity characterization of ordered sets, *Proc. 10th
     SE Conf. Comb., Graph Th., and Computing*, 529-540.

33. R. E. Jamison, Copoints in antimatroids, *Proc. 11th SE Conf. Comb.,
     Graph Th., and Computing, Congressus Numerantium*, 29(1980), 535-544.

34. R. E. Jamison, Hyperplane coverings and coaffine convexity over finite
     fields, *Math. Annalen*, 30(1978), 3-10.

35. R. E. Jamison, On convexity and the geometry of hyperplane complements,
     Bull, Soc. Math. Belgique 30(1978), 3-10.

36. R. E. Jamison, Partition numbers for trees and ordered sets, *Pac. J.
     Math.*, to be published.

37. R. E. Jamison, Convexity in graphs, in preparation.

38. R. E. Jamison, Antimatroids, in preparation.

39. R. E. Jamison, Combinatorial convexity in semilattices, in preparation.

40. B. Jonsson, Lattice-theoretic approach to projective and affine geometry,
     The Axiomatic Method, L. Henkin, P. Suppes, Alfred Tarski, etc., North-
     Holland Publ., Amsterdam, 1959, 188-203.

41. D. C. Kay and Womble, E. W., Axiomatic convexity theory and the relationship between the Carathéodory, Helly, and Radon numbers, *Pac. J. Math.* 38(1971), 471-485.

42. G. Koethe, *Topologische Lineare Raeume* I, Springer-Verlag, Berlin, 1960.

43. C. G. Lekkerkerker, and Boland, J. C., Representation of a finite graph by a set of intervals on the real line, *Fund. Math. Polska Akad. Nauk* 51(1962), 45.

44. B. Lindstrom, A theorem on families of sets, *J. Comb. Theory* (A) 13 (1972), 274-277.

45. E. H. Moore, *Introduction to a Form of General Analysis*, New Haven Math. Colloquium, Yale Univ. Press, New Haven, 1910.

46. L. Nachbin, *Topology and Order*, Van Nostrand Math. Studies #4, Van Nostrand, Princeton, 1965.

47. W. D. Neumann, On the quasivariety of convex subsets of affine spaces, *Archiv d. Math.* 21(1970), 11-16.

48. J. R. Reay, Carathéodory theorems in convex product structures, *Pac. J. Math.* 35(1970), 227-256.

49. H. E. Scarf, An observation on the structure of production sets with indivisibilities, *Proc. Natl. Acad. Sci. USA* 74(1977), 3637-3641.

50. J. Schmidt, Ueber die Rolle der transfiniten Schlussweisen in einer allgemeinen Ideal-theorie, *Math. Nachr.* 7(1952), 165-182.

51. J. Schmidt, Einige grundlegende Begriffe und Saetze aus der Theorie der Huellen-operatoren, *Bericht ueber die Mathematiker-Tagung in Berlin, Januar 1953*, Deutscher Verlag der Wissenschaften, Berlin, 1953, 21-48.

52. G. Sierksma, Caratheodory and Helly numbers of convex product structures, *Pac. J. Math.* 61(1975), 275-282.

53. G. Sierksma, and Reay, J. R., A Tverberg-type generalization of the Helly number of a convexity space, Preprint, Western Washington Univ., 1979.

54. A. Tarski, Fundamentale Begriffe der Methodologie der deduktiven Wissenschaften, *Monatsh. Math. Phys.* 37(1930), 360-404.

55. W. T. Trotter, and Moore, J. I., Characterization problems for graphs, partially ordered sets, lattices, and families of sets, *Discrete Math.* 16(1976), 361-381.

56. W. T. Tutte, A homotopy theorem for matroids I, II, *Trans. AMS* 88(1958), 144-174.

57. W. T. Tutte, Matroids and graphs, *Trans. AMS* 90(1959), 527-552.

58. W. T. Tutte, *Introduction to the Theory of Matroids*, American Elsevier, New York, 1971.

59. H. Tverberg, On equal unions of sets, *Studies in Pure Math.*, L. Mirski, ed., Academic Press, London (1971), 249-250.

60. M. van de Vel, Pseudo-boundaries and pseudo-interiors for topological convexities I, II, *Rapport Nr. 100*, Vrjie Universiteit Amsterdam, 1979.

61. H. van Maaren, Pseudo-ordered fields, *Indag. Math.* 36(1974), 463-476.

62. H. van Maaren, *Algebraic Simplices in Modules: On Concepts of Closure and Generalizations of Convexity*, Dissertation, Rijksuniversiteit Utrecht, 1977.

63. D. J. A. Welsh, *Matroid Theory*, Academic Press, London, 1976.

64. H. Whitney, On the abstract properties of linear dependence, *Amer. J. Math.* 57(1935), 507-533.

65. R. Wille, *Kongruenzklassengeometrien*, Lecture Notes in Math. #113, Springer-Verlag, Berlin, 1970.

66. R. M. Wilson, An existence theory for pairwise balanced designs, I, *J. Comb. Th.* (A) 13(1972), 220-245.

67. O. Zariski, and Samuel, P., *Commutative Algebra*, I, Van Nostrand, Princeton, 1958.

68. J. Kahn, and Kung, J., *Bull. AMS*, 3(1980), 857-858.

NOTE BY EDITOR. Problem 1 mentioned in this paper has been partially solved, although various solutions were known previously in unpublished manuscripts. A little known paper by J. P. Doignon [Caractérisations d'espaces de Pasch-Peano, *Acad. Roy. Belg. Cl. Sci.* (5) 62(1976), 679-699] deals with the most difficult aspect of embedding an alignment of dimension $\geqslant 2$ with certain properties (A, B, C we shall call them) in a convex subset of a real vector space. A paper by J. H. M. Whitfield and S. Yong [A characterization of line spaces, *Can. Math. Bull.*, 24 No. 3(1981), 351-357] then shows that an alignment satisfying very natural convexity properties of R^n (denoted JHC, REG, STR and CMP, shown to be independent in the paper) also possesses the properties A, B and C and thus, by the previous result, is isomorphic to an aligned subspace of a real vector space. Doignon's theorem was established for dimension $\geqslant 3$ independently by J. Cantwell and this editor [Geometric convexity. III: Embedding, *Trans. AMS* 246(1978), 211-230] using conventional methods in the foundations of geometry; Doignon's argument makes use of a 1938 theorem of Sperner.

OPEN PROBLEMS AROUND RADON'S THEOREM

John R. Reay

Department of Mathematics and Computer Science
Western Washington University
Bellingham, Washington

1. INTRODUCTION

The well-known theorem of Radon [1921] asserts that

THEOREM 1.1   Each set of $d + 2$ points in R^d is the union of two disjoint subsets whose convex hulls have a common point.

The purpose of this lecture is to list a number of open problems and conjectures that are related to this classical theorem of Radon, and identify some of the recent work that generated or partially solved these problems. No proofs will be given, and no attempt is made at making this a complete survey. For good surveys of the area and its bibliography, see the essays of Danzer-Grünbaum-Klee [1963] (for early literature), Doignon-Valette [1975], and Eckhoff [1979], and also the expository article of Peterson [1972].

Much of the work of the last decade related to Radon's Theorem has been in an abstract setting, but we deliberately leave this until last since (a) the familiar R^d is certainly the best place to begin and is the source model for most axiomatic treatments, (b) Gerard Sierksma will cover the most recent abstract results in the next lecture, and (c) the open problems and results in R^d should always be kept in mind when considering abstract convexity.

We will first consider independence and (m,k)-divisible sets, them primitive Radon partitions and convex polytopes, and finally the alignment (abstract convexity) setting.

2. INDEPENDENCE IN (m,k)-DIVISIBLE SETS

A set $S \subset R^d$ is *(m,k)-divisible* if it may be partitioned into m (pairwise disjoint) subsets whose convex hulls intersect in a set of dimension at least k. Such a partitioning of S is called an *(m,k)-partition*. (The empty set has dimension -1; assume $2 \leqslant m$ and $0 \leqslant k \leqslant d$.) An (m,0)-divisible set is also called *m-divisible*. Thus Radon's Theorem asserts that each (d + 2)-set in R^d is 2-divisible. Birch [1959] conjectured and Tverberg [1966] proved the following

THEOREM 2.1 Each [(m - 1)(d + 1) + 1]-set in R^d is m-divisible.

Although Tverberg's proof made use of *algebraic independence* of sets (i.e., the t points of S have t·d algebraically independent real coordinates over the field of rationals), his proof can be modified to use the weaker condition of *strong independence* which essentially asserts that the points of S do not form pencils of lines, or books of planes, etc. Set $S \subset R^d$ is *strongly independent* provided each finite collection $S_1, \ldots, S_t$ of disjoint subsets has the property: if card $S_i = d_i + 1 \leqslant d + 1$ then $\dim(\cap_{i=1}^{t} \text{aff } S_i)$ $= \max\{-1, d - \sum_{i=1}^{t} (d - d_i)\}$. That is, for any appropriate collection of flats, the codimension of the intersection is the sum of the codimensions. Thus strong independence of a set S clearly implies general position.

For a set to be (m,k)-divisible with $k \geqslant 2$ some independence is necessary to avoid having all the points on a line. It is an open problem to determine this best (weakest) independence.

PROBLEM 1. Conjecture: Each [(m - 1)(d + 1) + k + 1]-set in general position in R^d is (m,k)-divisible.

Peterson [1972] gives a geometric proof in case m = 2; Reay [1968] proves the case d = 2; and Reay [1979] establishes the case d = m = 3. If the

FIGURE 1

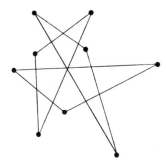

hypothesis of general position is strengthened to strong independence then
the conjecture is true as a corollary of the following result by Reay [1968]:

THEOREM 2.2   If the $[(m - 1)(d + 1) + k + 1]$-set $S \subset R^d$ is strongly independent and $S = S_1 \cup \cdots \cup S_m$ is a partition with each card $S_i \leqslant d + 1$, then
if $\cap_{i=1}^{m}$ conv S_i is nonempty, it is of dimension k.

The example shown in Fig. 1, where $d = k = 2$, $m = 3$, consists of nine
points in general postion which are (3,2)-divisible even though the indicated partition shows that 2.2 fails. Thus, strong independence cannot be replaced by general position.

A cheap generalization of Radon's Theorem is obtained by replacing the
field of real numbers by any ordered division ring, since the usual proof
(Radon's original) remains valid. But Tverberg's proof of 2.1 relies in
several places on properties of the real numbers and does not have such a
trivial generalization. Doignon and Valette [1977] show that 2.1 does remain valid in any d-dimensional affine space A^d over an ordered division
ring (i.e., skew-field), and give a more general proof of Tverberg's Theorem.
It turns out that the natural independence condition to consider in this
setting is *full independence*, which is defined in the same way as strong
independence except that aff S_i is replaced by proj S_i, the projective subspace generated by S_i in the projective extension of A^d. Surprisingly,
strong independence implies full independence only in spaces of dimension
less than five, while the four vertices of a parallelogram in any affine
space show that full independence does not imply strong independence. In
fact, Doignon and Valette [1977] show that there exist finite strongly

independent sets which are maximal under inclusion, and characterize them
as the sets which are strongly independent but not fully independent! (See
also Doignon-Valette [1975].) Eckhoff [1979] remarks that 2.2 remains valid
with full instead of strong independence.

PROBLEM 2. Is there a better general setting in which both full and strong
independence become extensions of a more general (weaker) independence?

Next note that no independence at all is required in the special case
$k = 1$, that is, when we wish to partition a set S into subsets whose convex
hulls have a common line segment. However, only crude upper bounds are known,
in general, on the cardinality of S necessary for $(m,1)$-divisibility. Reay
[1979-b] shows that each $(2(d + 1)(m - 1) + 1)$-set in R^d is $(m,1)$-divisible.
A much more reasonable conjecture about the correct cardinality is the following.

PROBLEM 3. Conjecture: Each $(2d(m - 1) + 2)$-set in R^d has a $((d + 1)(m - 1)
+ 2)$-subset which is $(m,1)$-divisible.

Easy examples show that the number $2d(m - 1) + 2$ cannot be reduced;
it is clearly true if $d = 1$, and it was proved for $d = 2$ by Reay [1979-b]
and for $m = 2$ by Eckhoff [1976]. A related question of Doignon which requires no independence condition is

PROBLEM 4. Conjecture: Each $(2d + 1)$-set $S \subset R^d$ has a Radon partition
$S = S_1 \cup S_2$ such that $\dim(\text{conv } S_1 \cap \text{conv } S_2) \geqslant \min\{\dim S_1, \dim S_2\}$.

Attempts to establish the conjecture in Problem 3 produced the following notation which has been useful in abstract convexity. Point x is an
m-divisible point of set $S \subset R^d$ if $x \in \cap_{i=1}^{m} \text{conv } S_i$ for some partition $S =
S_1 \cup \cdots \cup S_m$ of S. We denote the set of m-divisible points of S by $D_m(S)$.
The next result seems to support the conjecture in Problem 3 (see Reay [1979-b]):

THEOREM 2.3 Each $(2d(m - 1) + 2)$-set in R^d has at least two distinct m-
divisible points.

Also let $C_m(S)$, called the *m-core of S*, denote the set of all points $y \in R^d$

so that each closed half-space containing y also contains at least m points
of S. The m-core of S may equivalently be defined in a way which generalizes
to an abstract convexity space:

$$C_m(S) = \cap\{\text{conv } T \mid T \subseteq S \text{ and } \text{card}(S \sim T) \leq m - 1\}$$

It would certainly be of interest to know under what conditions the set $D_m(S)$
of m-divisible points is a convex set, since 2.3 then implies that the conjecture in Problem 3 is true.

PROBLEM 5.   (a) When is $D_m(S)$ convex?

   (b) Conjecture: If $S \subseteq R^d$ is finite, so $C_m(S)$ is a polytope, then
        the vertices of $C_m(S)$ lie in $D_m(S)$, or equivalently, $C_m(S) = $
        $\text{conv}(D_m(S))$.

   (c) Under what conditions is core $C_m(S)$ at least k-dimensional?

   It is clear that $D_m(S) \subseteq C_m(S)$, and the methods of Birch [1959], applied to planar sets S, show that $D_m(S) = C_m(S)$ whenever $3m <$ card S. Further results and open problems concerning the m-core $C_m(S)$ will be given by
Sierksma.

   A possible first step towards solving Problem 2 and unifying some of
the above results has been made by Doignon [1980]. His first result shows
that the general position of S (any subset of at most d + 1 points is affinely independent) is too strong a condition in Problem 1 in case m = 2
and $k \geq 2$.   (The example ┄┼┄ in R^2 shows why $k \geq 2$ is necessary.)

THEOREM 2.3A.   (a) A (d + k + 2)-set $S \subseteq R^d$ is (2,k)-divisible for $2 \leq k$
$\leq d$ provided each d + 2 points of S affinely span R^d. (It is not necessary
that each d + 1 points affinely span R^d.) (b) More generally $S \subseteq R^d$ is
(2,k)-divisible if card $S \geq d + k + 2$, k > 2 and $S \sim T$ affinely spans R^d
whenever $2 \leq$ card $T \leq k$. (Combining this result and the known cases of Problem 1 gives the next even more general result.)   (c) A set $S \subseteq R^d$ of at
least $\alpha(k,d) = (d - k + 2)k + 2$ points is (2,k)-divisible provided no 2k + 2
of them lie in a k-dimensional flat, and $2 \leq k$.

PROBLEM 6.   (a) Can the bounds $\alpha(k,d)$ given in 2.3A be lowered?

The bound $\alpha(k,d) = (d - k + 2)k + 2$ is best if $k = 2$ or $d - 1$ or d, but Problem 6 is open otherwise. We say a set is *k-independent* if any $k + 1$ or fewer of its points are affinely independent. Thus k-independence is weaker than general position unless $k = d$, in which case they are equivalent. Eckhoff has commented that 2.3A implies the following:

THEOREM 2.4. A set $S \subset R^d$ of at least $\beta(k,d) = (d - k + 2)(k + 1)$ points is (2,k)-divisible if it is k-independent.

Again the bound on card S is best if $k = 0$ or 1 or d.

PROBLEM 6. (b) Can the bounds $\beta(k,d)$ given in 2.4 be lowered?

It would be nice to know how the above two results can be extended to (m,k)-divisible sets; Problem 4 might be the key to that project.

PROBLEM 7. Conjecture: Any k-independent set $S \subset R^d$ of at least $(m - 1) \cdot (2d - k + 1) + k + 1$ points is (m,k)-divisible $(k > 0)$. (In fact, any such S would have a $[(m - 1)(d + 1) + k + 1]$-subset which is (m,k)-divisible.)

Note that if $k = d$ this reduces to a case of the conjecture in Problem 1, while if $k = 1$ so that 1-independence just implies distinct points, this becomes Problem 3. The standard example, which shows that the number $2d \cdot (m - 1) + 2$ in Problem 3 cannot be reduced, is $S = \{\alpha b \mid b \in B, \alpha = 0, \pm 1, \ldots, \pm(m - 1)\}$ where B is any linear basis for R^d. This set S is neither (m,1)-divisible nor 2-independent. Thus it is tempting to conjecture a stronger version of Problem 3 when S is 2-independent.

PROBLEM 8. A 2-independent $(2d(m - 1) + 1)$-set in R^d is (m,1)-divisible.

It seems unlikely that the cardinality constraint in Problem 8 is the best possible.

Tverberg [1968] extended 2.1 as follows

THEOREM 2.5. Let $m \geqslant 2$, $0 \leqslant c \leqslant d$, $s = (m - 1)(c + 1) + 1$, and let $F_1, F_2, \ldots, F_s$ be flats of codimension c in R^d. Then $\{1, \ldots, s\}$ has an m-partition $A_1, \ldots, A_m$ so that $\cap_{i=1}^{m} (\text{conv } \cup_{j \in A_i} F_j)$ contains a flat of codimension c.

This result reduces to 2.1 when $c = d$. By choosing all the flats F_i to be parallel in R^d when $c < d$ it is easy to see from 2.1 that the bound on S is the best possible. But the effect of making the flats F_i independent has been investigated only for the case where no algebraic condition exists. See Tverberg [1968] and Iversland [1969, Chapter 4].

PROBLEM 9.   (a)   What independence conditions on the flats F_i of 2.3 will allow the codimension c of the intersection to be reduced?

(b)   By increasing the number of flats by k ($1 \leqslant k \leqslant c$), what independence condition will assure the intersection in 2.5 is dimension at least $d - c + k$?

(c)   What results analogous to 2.5 are possible if the flats F_i are allowed to have different dimensions?

Robert Jamison has suggested a different variant of 2.1 and (m,k)-divisibility (S is (m,k)-divisible if there is a partition $S = S_1 \cup \cdots \cup S_m$ with $\cap_{i=1}^m$ conv S_i k-dimensional). Suppose we only require conv $S_i \cap$ conv S_j to be k-dimensional for each pair (i,j), $1 \leqslant i < j \leqslant m$. Perhaps a smaller cardinality for S would suffice for this weaker condition.

PROBLEM 10.   (a)   Each set $S \subset R^d$ of at least $T(d,m,n,0) = (m - 1)(d + 1) + 1$ points has an m-partition $S = S_1 \cup \cdots \cup S_m$ so that each n of the sets {conv S_i | $i = 1,\ldots,m$} has a nonempty (at least 0-dimensional) intersection. This is a weaker form of 2.1.

Conjecture: If $2 \leqslant n \leqslant m$, the number $T(d,m,n,0)$ given above is the best possible.

Clearly the case $n = m$ is just 2.1. The conjecture states that the cardinality condition of Tverberg's theorem cannot be improved in general when the intersection requirement is weakened ($m > n$). The conjecture follows directly from Helly's Theorem if $n \geqslant d + 1$, and is therefore trivial if $d = 1$. It is shown in Reay [1979-b] that the conjecture is true if $d = 2$ or if $d = m = 3$ and lower bounds on $T(d,m,n,0)$ are given.

We may define $T(d,m,n,k)$ similarly, where k denotes the minimal dimension of the intersection of any n of the m sets {conv S_i}. When $k > 1$ an additional independence condition on S is again necessary.

PROBLEM 10.   (b) Investigate the various independence conditions in determining the minimal value of $T(d,m,n,k)$.

(c) Conjecture: If $S \subset R^d$ is k-independent and $n \geqslant 2$, then $T(d,m,n,k)$
$= (m - 1)(d + 1) + k + 1$ (and thus T is independent of n).

If an (m,k)-divisible set has an abundance of points when forming an
(m,k)-partition, then it would be reasonable to distribute the excess as
uniformly as possible among the m subsets of the partition. just in case some
of the points get "stolen" or "lost" in the future. Let $L(d,v)$ be the smallest integer so that each set $S \subset R^d$ of at least $L(d,v)$ points has a $(2,0)$-
partition $S = S_1 \cup S_2$ with the property that $\cap_{i=1}^{2} \mathrm{conv}(S_i \sim T) \neq \phi$ whenever
any subset $T \subset S$ which is "stolen" has cardinality at most v.

PROBLEM 11.   (a) Determine $L(d,v)$.   [$L(d,0) = d + 2$ is just Radon's Theorem.]

(b) Conjecture: $L(d,v) = (v + 1)(d + 1) + 1$.

Example: To see that the conjecture is true when $d = 1$, consider $2v + 3$
points on a line alternately in S_1 and S_2. This number is best, because any
partition $S_1 \cup S_2$ of $2v + 2$ points would need exactly $v + 1$ in each S_i, yet
taking all v from one set leaving only the last point on the line in S_1, say,
would make $\cap_{i=1}^{2} \mathrm{conv}(S_i \sim T)$ empty.

The hard part of the conjecture, namely $L(d,v) \leqslant (v + 1)(d + 1) + 1$,
was proved by Larman [1972] in case $v = 1$ and by Strangeland [1978] for higher values of v. So "all that is needed" are the counter examples of $[(v + 1)$
$(d + 1)]$-sets in R^d, no $(2,0)$-partition of which can afford to have v points
stolen.

It was natural to extend the definition of $L(d,v)$ from $(2,0)$-partitions
to (m,k)-partitions. Let $L(d,m,k,v)$ denote the smallest integer so that
each k-independent set $S \subset R^d$ of at least $L(d,m,k,v)$ points has an (m,k)-
partition with the property that $\dim(\cap_{i=1}^{m} \mathrm{conv}(S_i \sim T)) \geqslant k$ whenever $T \subset S$
is any subset of at most v points. (Here $k \leqslant d$ and $m \geqslant 2$.) Clearly $L(d,m,0,0)$
is determined by Tverberg's Theorem 2.1 and $L(d,2,0,v)$ is Larman's function
given above. The following conjecture is due to Strangeland [1978].

PROBLEM 11.   (c) Determine $L(d,m,k,v)$, or determine L for the more restric-
tive class of strongly independent sets in R^d.

(d) Conjecture: $L(d,m,0,v) = (m - 1)d(v + 1) + m + v$.

See Eckhoff [1979] for related results and conjectures, some of which relate to the function T of Problem 10, and for the related problems concerning a common transversal which meets each of the m (possibly disjoint) convex hulls formed by a partition of set S. We state only the following (Iversland [1969]).

PROBLEM 12.   Does every $[(m - 1)d + n + 1]$-set $S \subset R^d$ have a partition $S = \cup_{i=1}^{m} S_i$ so that among the convex hulls $\{conv \, S_i\}$ there are n, each of which intersect all m convex hulls?

Let S be a $(d + 2)$-set in R^d, and let $S = S_1 \cup S_2$ be the partition guaranteed by Radon's Theorem. It is well-known that the partition is unique if and only if S is in general position. (See Peterson [1972] for a geometric proof.) Further, this $(2,0)$-partition gives a unique single point as the intersection of the convex hulls. There is no known analogous result for $(m,0)$-partitions of a set S if $m \geqslant 3$.

PROBLEM 13.   (a) For which sets S is it true that every (m,k)-partition of S gives the same (unique) k-dimensional polytope for the intersection of the corresponding convex hulls?

(b)   Which (m,k)-divisible sets S have a unique (m,k)-partition?

If a set S has an $(m,0)$-partition then it is unique only if no proper subset has an $(m,0)$-partition, so we consider only such sets.

Example.   The set shown in R^3 (Fig. 2) has a unique $(3,0)$-partition $\{(0,0,2),(0,0,-2)\} \cup \{(0,2,0),(0,-2,0)\} \cup \{(-1,0,1),(-1,0,-1),(1,-1,0),(1,1,0)\}$. Since $C_3(S) = \{(0,0,0)\} = D_3(S)$, the origin is the only 3-divisible point. Yet this set of $(n - 1)(d + 1) = 8$ points is not in general position. On the other hand, *any* $(3,0)$-divisible set of at least 9 points appears to have many $(3,0)$-partitions. A lower bound has been conjectured by Sierksma.

PROBLEM 14.   (a)   (Dutch Cheese Problem) Conjecture: If $S \subset R^d$ is an $[(m - 1)(d + 1) + 1]$-set which is $(m,0)$-divisible, but no proper subset of S is $(m,0)$-divisible, then S has at least $[(m - 1)!]^d$ distinct $(m,0)$-partitions.

(b)   If $S \subset R^d$ is a strongly independent $[(m - 1)(d + 1) + k + 1]$-set, what is the minimum number of distinct (m,k)-partitions?

FIGURE 2

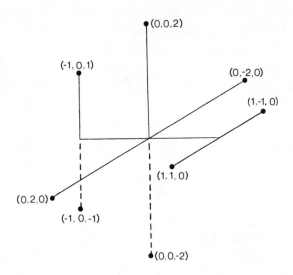

The bound $[(m - 1)!]^d$ is based on the fact that $d + 1$ independent points, each of multiplicity $m - 1$, together with a single point interior to this simplex, forms a $[(m - 1)(d + 1) + 1]$-set of (nondistinct) points with this number of distinct $(m,0)$-partitions. The name of the conjecture comes from the fact that Sierksma [1979] offers a Dutch cheese for the first solution. If $m = 2$, so that the bound reduces to 1, and card $S = d + 2$, the problem reduces to the well-known result quoted above.

3. CONVEX POLYTOPES AND PRIMITIVE RADON PARTITIONS

We bridge the jump from the last section with this problem posed by Tverberg in the 1978 Oberwolfach conference on convexity.

PROBLEM 15. Conjecture: If $P \subset R^{(m-1)(d+1)}$ is a (nondegenerate) convex polytope, and $\pi: \text{bd } P \to R^d$ is a continuous projection then there exist pairwise disjoint faces $\{F_i\}_{i=1}^m$ of P so that $\cap_{i=1}^m \pi(F_i) \neq \phi$.

Since any $[(m - 1)(d + 1) + 1]$-set in R^d is $\pi(T)$ for some simplex T in $R^{(m-1)(d+1)}$, this conjecture would give an alternate proof of 2.1. The case $m = 2$ was proved by Bajmóczy and Bárány [1979] (giving an alternate proof of Radon's Theorem). More generally, the conjecture was proved by Bárány, Shlosman, and Szücs for all cases where m is a prime!

The following problem was posed by Peter McMullen.

PROBLEM 16. Is 2d + 1 the maximum number of points in general position in R^d (but otherwise arbitrarily placed) which can be mapped, by a permissible projective transformation, onto the vertices of a polytope in R^d?

At first glance this problem has little to do with Radon's Theorem. But by use of Gale diagrams it is a reformulation of one case of Problem 11(b) above. (See Larman [1972].) In fact, Gale diagrams have played a role in a number of problems around Radon's Theorem. For example, Shephard [1969] has characterized the unique Radon partitions $\{S_1, S_2\}$ of a $(d + 2)$-set S in general position in R^d by the fact that the Gale transforms $\bar{S}_1$ and $\bar{S}_2$ lie on opposite sides of the origin in the transform space $R^{(d+2)-d-1} = R^1$. More generally, he characterizes the primitive Radon partitions of a finite set in general position (definitions below) in terms of Gale transforms, and Doignon [1980] has recently characterized those Radon partitions $\{S_1, S_2\}$ of $S \subset R^d$ for which

$$\dim(\text{conv } S_1 \cap \text{conv } S_2) \leqslant k$$

in terms of Gale transforms. Peter McMullen [1979] has a survey report on the various applications to which Gale transforms have been put. Also see Doignon-Valette [1975].

PROBLEM 17. Give characterizations, in terms of the Gale transform of a finite set $S \subset R^d$, of

    (a)   the partitions $\{S_1, \ldots, S_m\}$ of S which are (m,k)-partitions (S in general position and $m \geqslant 3$).

    (b)   The (m,k)-partitions of S when S has another independence, such as k-independence or strong independence ($m \geqslant 2$).

Most of the remaining results and problems of this section are stated for (2,0)-partitions, called *Radon* partitions, of sets S. However, they generally could be rephrased in the setting of (m,k)-partitions. The sets S_i of a Radon partition $S = S_1 \cup S_2$ are called *components* of the partition. $\{T_1, T_2\}$ is a *Radon partition in* S (as opposed to *of* S) if it is a Radon partition and if $T_1 \cup T_2$ is a subset of S. The partition $\{S_1, S_2\}$ *extends* $\{T_1, T_2\}$

if $T_i \subset S_i$. $\{T_1, T_2\}$ is a *primitive* Radon partition in S if it is minimal in the sense that it does not extend any distinct Radon partition in S. Two sets S and T are said to be *Radon equivalent* provided there exists a bijection f : S → T which preserves Radon partitions, or equivalently a bijection g : S → T which preserves primitive Radon partitions. The equivalence classes induced by this equivalence relation are called *Radon types*. See Eckhoff [1974, 1975, 1979]. These definitions will be thought of in terms of sets S, T $\subset R^d$, but these terms along with simplex, general position, faces, k-neighborly, and most of the following results and problems can be formulated for sets S and T in any real linear space (finite or infinite dimensional). See Petty [1975]. We collect some well-known facts about primitive partitions in:

THEOREM 3.1.   (a) Each Radon partition extends a primitive one.

(b) For each primitive partition $\{A, B\}$ in S

(1) $A \cup B$ is in general position in $aff(A \cup B)$.

(2) $card(A \cup B) = dim(A \cup B) + 2$, is finite, and is $\leq d + 2$ if $S \subset R^d$.

(3) $A \cup B$ is (2,k)-divisible iff k = 0, and its Radon partition is unique.

(4) points x,y $\in A \cup B$ lie in different components iff $aff(A \cup B \sim \{x,y\})$ separates x and y in $aff(A \cup B)$.

(c) If $S \subset R^d$ has d + 2 general position points, not all on a common (d − 1)-sphere, and $S = A \cup B$ is its unique primitive partition, then

$$A = \{x \in S \mid x \text{ lies inside the } (d-1)\text{-sphere through } S \sim x\}$$

and

$$B = \{x \in S \mid x \text{ lies outside the } (d-1)\text{-sphere through } S \sim x\}$$

Primitive Radon partitions were first used by Hare and Kennelly [1971] in an attempt to discover when certain points of a set S lay in the same component of a Radon partition. They obtained part (c) of 3.1 above and the following for the special case when S is a finite set in general position and in R^d. The present form is due independently to Petty [1975] and Doignon-Valette [1975].

THEOREM 3.2. A subset $T \subset S$ lies in a component of some Radon partition of S if and only if

(a) S ~ T is affinely dependent, or

(b) conv T ∩ aff(S ~ T) ≠ φ

PROBLEM 18. (a) When does a subset T ⊆ S lie in one (or in all but one)
component of a (m,0)-partition of S?   (m ⩾ 3)

(b) When does a subset T ⊆ S necessarily meet every (or at least two)
component of a (m,0)-partition of S?   (m ⩾ 2)

In general, it seems to be very difficult to say when two finite sets
in R^d are Radon equivalent.

PROBLEM 19. Characterize the Radon types of finite sets of points in R^d.

Progress has been made only for certain types of sets. Breen [1973]
implicitly characterized the Radon types for the vertex sets of cyclic polytopes. Eckhoff [1975] used Gale diagrams to classify the Radon types of
general position sets with at most d + 3 points. Any two such sets, which
are vertex sets of d-polytopes, are Radon equivalent if and only if the polytopes have the same combinatorial type. More generally, Breen [1972] showed
that the combinatorial type of any polytope is determined by the Radon partitions of the vertex set, and in fact, by the primitive Radon partitions
which make use of a (fixed but arbitrarily chosen) vertex v. However, easy
examples show that combinatorial equivalence of polytopes does not imply
Radon equivalence of their vertex sets. The key result of Breen was the
following characterization of the facial structure of d-polytopes.

THEOREM 3.3 Let S be the vertex set of a d-polytope and F ⊂ S. The set
conv F is a face of conv S if and only if T_1 ⊂ F ⇒ T_2 ⊂ F for every primitive Radon partition {T_1, T_2} in S. In particular F determines a face of a
simplicial polytope conv S if and only if no subset of F is a component of a
Radon partition of S.

PROBLEM 20. (a) Given a finite set S and a minimal collection C of partitions in S (minimal in the sense that no partition in the collection extends
another), when does there exist a polytope in R^d whose vertices correspond
to S and whose primitive Radon partitions correspond to the partitions of C?

(b) If such a polytope exists, what can be said about the range of d?

See Bland and Las Vergnas [1978] for a statement of this problem in the setting of orientable matroids.

There have been a number of results about the number $N(S)$ of distinct Radon partitions that a set $S \subset R^d$ may have. See Eckhoff [1979]. One of the most appealing is the following due to Winder [1966].

THEOREM 3.4. $N(S)$ is the number of subsets $T \subset S$ for which dim T + card T is even.

PROBLEM 21. What is a "natural" correspondence between the Radon partitions of S and the subsets $T \subset S$ with dim T + card T even?

Theorem 3.4 was not stated by Winder in the form given above; in fact, Winder never mentions Radon's Theorem at all. The connection with Winder's work lies in a definition of Brylawski [1976], who lets $R(S)$ be the number of subsets of $S \subset R^d$ which can be separated from their complements by a hyperplane. Radon's Theorem then becomes: $R(S) < 2^d$ if card $S \geqslant d + 2$.

There are a number of results which give an upper bound on $N(S)$ for various classes of sets S. (See Brylawski [1976, 1977].) For lower bounds it is natural to consider primitive partitions of S.

PROBLEM 22. (a) Determine the smallest number of primitive Radon partitions of a set of s points in R^d. For which set is the lower bound attained?

(b) Determine bounds on the number of distinct (m,k)-partitions of a set $S \subset R^d$ in terms of card S and the various independence conditions that may be imposed on S.

This problem should be compared with the Dutch Cheese Problem (Problem 14). See Lindquist-Sierksma [1980] and Sierksma [1980] for related results in an abstract convexity structure. See Doignon [1980b] for a setting of the problem using matroids, a general conjecture, and proofs of certain special cases.

It has long been an open problem to determine which vectors $f = (f_0, f_1, \ldots, f_{d-1})$ of natural numbers represent the f-vectors of a d-polytope if $d \geqslant 4$. (See Barnette-Reay [1973].) A similar problem arises when we study the sizes of the components in Radon partitions of a set $S \subset R^d$. A partition $\{T_1, T_2\}$ in S is of *type* $\{t_1, t_2\}$ if card $T_i = t_i$. For a set $S \subset R^d$ of s points, let

$r_i = r_i(S)$ be the number of Radon partitions of S of type $\{i, s - i\}$ and let
$P_i = P_i(S)$ be the number of primitive partitions in S of type $\{i, d + 2 - i\}$.
Then the vectors $(r_1, \ldots, r_{[s/2]})$ and $(P_1, \ldots, P_{[d/2]+1})$ are called the *Radon vector* and the *primitive Radon vector* of S, respectively.

PROBLEM 23. (a) Which vectors $(r_1, \ldots, r_{[s/2]})$ and $(P_1, \ldots, P_{[d/2]+1})$ of natural numbers represent the Radon vector and primitive Radon vector, respectively, of some set S of s points in general position in R^d?

(b) Find upper or lower bounds on r_i or P_i for sets of s general position points in R^d.

Except for certain cases this appears to be a very hard problem. The analogue of the Euler hyperplane for f-vectors are the hyperplanes

$$\sum_{i=1}^{[d/2]+1} P_i = \binom{s}{d+2} \tag{3.4}$$

and

$$\sum_{i=1}^{[s/2]} r_i = N(S) \tag{3.5}$$

Equation (3.4) is the only linear relation on the coordinates of every primitive Radon vector, but other linear relations than (3.5) hold for the Radon vectors.

If $s = d + 2$, the uniqueness of the (primitive) Radon partition of S makes the characterization trivial; all coordinates are zero except for one whose value of one could appear in any position.

If $s \geq d + 3$, then the Radon vector and the primitive Radon vector of S will have an initial segment of zeros as coordinates and thereafter will always be positive. Furthermore, the vertices of a k-neighborly polytope which is not $(k + 1)$-neighborly will have a vector whose initial segment of zeros is of length exactly k. This was proved for Radon vectors by Shephard [1969] and for primitive Radon vectors by Eckhoff [1974] using Gale transforms. When $s = d + 3$, so that the Gale diagrams are two-dimensional, Eckhoff also gave a complete characterization of the Radon and primitive Radon vectors of S, and showed that either vector determined the other.

If $s = d + 4$, then knowing either the Radon vector or the primitive Radon vector of a set S does not allow us to determine the other vector. But here the two vectors are related through the f-vector of conv S, as was shown by Kramer [1975]:

THEOREM 3.6.   If $S \subset R^d$ is in general position, card $S = s = d + 4$, and f is
the f-vector of conv S, then

$$r_i = p_{i-1} - \sum_{j=0}^{i} (-1)^{i-j} \binom{s-j}{i-j} f_{j-1} \qquad \text{when } i = 1,\ldots,[d/2] + 1$$

Theorem 3.6 will not hold if s becomes large compared to d in general.

PROBLEM 24.   Characterize the sets S of at least $d + 3$ points for which 3.6
is true, i.e., primitive Radon vectors are determined by the Radon vector of
S and the f-vector of conv S.

Note that if Problem 24 were established, then the Dehn-Sommerville
equations for conv S would yield another linear relation between the r_i and
p_i.

PROBLEM 25.   Find a complete system of linear relations which are satisfied
by the Radon vector and the primitive Radon vector.

Several such linear relations have been found by Kramer [1975].   The
necessary calculations make fundamental use of Breen's [1973] characterization of the primitive Radon partitions of the cyclic polytope.
New proofs of Radon's Theorem continue to be found by investigators of
other areas of mathematics.   For example, Dudley [1979] supplies a proof
based on a series of papers with probabilistic implications.   We close this
section with the following probabilistic problem:

PROBLEM 26.   Let S be a set of $d + 2$ points chosen randomly in a bounded
open domain $D \subset R^d$.   Find the probability P_i that the unique primitive Radon
partition of S is of type $T_i = \{i, d + 2 - i\}$ for $i = 1,2,\ldots,[d/2] + 1$.

The probability vector $(P_1,\ldots,P_{[d/2]+1})$ obviously depends on the domain D, but this restriction could easily be replaced by requiring only that
the origin is one of the $d + 2$ points of S, and the other points are chosen
randomly in R^d and form a set of diameter at most one.   This problem was
considered in the plane long before Radon's time.   Sylvester (while studying calculus of variations) conjectured that a disk (the interior of a circle)
was the domain $D \subset R^2$ which minimized the probability of four randomly chosen

points from D having a Radon partition of type {1,3} (that is, one point lies in the convex hull of the other three). In a survey article written two years before Radon was born, Crofton [1885] shows this probability is $P_1 = 35/(12\pi^2)$ when $D \subset R^2$ is a disk. See Klee [1969] for a survey of more recent references and results.

4. ABSTRACT CONVEXITY SPACES

The early survey paper of Danzer-Grunbaum-Klee [1963] listed a number of ways in which the notion of convexity had been generalized. Radon's Theorem had been considered in many of these settings, but mostly as only one part of an investigation of a particular space or axiomatic structure. One suggestion of their paper was to "study the inter-relationships of Radon's, Helly's, and Carathéodory's properties in a system which had convex sets closed under intersection, and perhaps other axioms of convexity." (Definitions are given below.) A number of subsequent papers investigated this interrelationship in a specific special setting such as convex-product spaces, order convexities, real linear spaces, semilattices and lattices, spherical convexities, or trees, etc. More recently this interrelationship has been investigated in as general a setting as the definitions will allow, in the hope of getting results which are valid in most of the special settings. This latter effort was started by Kay and Womble [1971] and Jamison [1970] in particular, and carried on by Doignon-Valette [1975], Doignon-Reay-Sierksma [1979], R. Hammer [1977], Jamison [1979], Sierksma [1977], and others. One interesting side effect has been to turn the problem around. Instead of including Radon numbers, etc. in a study of some particular structure, theorems about Radon numbers, etc. are being proved in a general setting, and then examples of particular structures are being sought in an attempt to prove that the theorems are as sharp as possible.

We now turn to some definitions. The one common axiom is *property (I)*: the intersection of any collection of convex sets is a convex set. An *abstract convexity space* is any pair (X,C) where X is a set and C a collection of subsets (called convex sets) which satisfy $\phi \in C$, $X \in C$ and property (I). An *alignment* is an abstract convexity space with C closed under nested unions. The *C-hull* (generalized convex hull) of a set $S \subset X$ is the set $C(S) = \cap\{A \in C \mid S \subset A\}$. The *m-Radon number* $r(m)$ of (X,C) is the least positive integer so that each set $S \subset X$ has a Radon m-partition provided card $S \geqslant r(m)$. The number $r = r(2)$ is called the *Radon number*. The *n-Helly number* $h(n)$ of

(X,C) is the least nonnegative integer so that each set $S \subset X$ has $C_n(S) \neq \phi$
provided card $S \geq h(n) + 1$.   (See Section 2 above for a definition of $C_n(S)$,
the *n-core* of S.)   The number $h = h(2)$ is called the *Helly number* because
$h(2)$ may equivalently be defined as the least nonnegative number so that the
intersection of any finite collection of convex sets is nonempty provided
the intersection of each subcollection of at most $h(2)$ of its sets is nonempty.   The *exchange number* e of (X,C) is the least positive integer so that
for each $x \in X$ and each finite $S \subset X$, $C(S) \subset \cup_{y \in S} C(x \cup (S \sim y))$ provided
card $S \geq e$.   The *Caratheodory number* c of (X,C) is the least nonnegative
integer so that $C(S) = \cup \{C(T) \mid T \subset S,$ card $T \leq c\}$ for all $S \subset X$.

   We summarize various known results in the following theorem:

THEOREM 4.1.

   (a) $h(n) \leq r(n + 1) - 1$

   (b) $r \leq c(h - 1) + 2$

   (c) $r(n) \leq c((n - 1)h - 1) + 2$

   (d) $r(n) \leq ((n - 1)h - 1) \max\{h, e - 1\} + 2$

   (e) $e \leq c + 1 \leq \max\{h + 1, e\}$ in an alignment

   (f) If $r(m_0)$ is finite (exists) for some integer m_0 in a space (X,C),
        then $h(n)$ and $r(m)$ are finite for every allowable value of m and
        n.

   (g) If $h(n_0)$ is finite for some integer n_0, then $h(n)$ is finite for
        all (allowable) values of n.

Part (a) is called the *Levi relation* since Levi [1951] proved the special
case $n = 1$.   Part (b) is called the *Eckhoff-Jamison relation* (established
independently by them in about 1976).   It improves the earlier result $r \leq$
$ch + 1$ due to Kay-Womble [1971] and led to the generalizations (c) and (d).
See Sierksma [1976, 1977] for proofs of (b) and (e) and an extensive bibliography of results up to that time.   Part (c) is due in its present form to
Doignon, and part (f) is due to Jamison.   See Doignon-Reay-Sierksma [1979]
for further details and for proofs of several of these results.

   For part (f) of 4.1 various bounds have been suggested, some of which
are known to be sharp (e.g., lower bounds on $r(m)$ in terms of r).   As one
example, Eckhoff [1979] makes the following conjecture, which reduces to
Tverberg's Theorem 2.1 in R^d with usual convex sets.

PROBLEM 27. Conjecture: $r(m) \leqslant (m - 1)(r - 1) + 1$

This brief survey of axiomatic convexity has ignored many nice results relating to Radon's Theorem which appear in a variety of special settings. A survey in the same detail as the previous two section would likely produce a list of an additional 25 problems. Instead, we will summarize them in one last problem reminiscent of Danzer-Grunbaum-Klee's original question.

PROBLEM 28. Study the relations between the Carathéodory, exchange, (generalized) Radon, and Helly numbers, and determine in which abstract convexity spaces the general results are sharp and in which specific spaces better results can be obtained.

REFERENCES

1. E. Bajmóczy and I. Bárány, On a common generalization of Borsuk's and Radon's theorem, *Acta Math. Acad. Scientiarum Hungaricae* 34(3-4)(1979), 347-350.

2. I. Bárány, S. Shlosman, and A. Szücs, On a topological generalization of a theorem of Tverberg, to be published in *J. London Math. Soc.*

3. D. Barnette and J. Reay, Projections of f-vectors of four-polytopes, *J. Comb. Theory* 15(1973), 200-209.

4. B. Birch, On 3N points in a plane, *Proc. Cambridge Philos. Soc.* 55(1959), 289-293.

5. R. Bland and M. Las Vergnas, Orientability of Matroids, *J. Comb. Theory* (B) 24(1978), 94-123.

6. M. Breen, Determining a polytope by Radon partitions, *Pacific J. Math.* 43(1972), 27-37.

7. M. Breen, Primitive Radon partitions for cyclic polytopes, *Israel J. Math.* 15(1973), 156-157.

8. T. Brylawski, A combinatorial perspective on the Radon convexity theorem, *Geometriae Dedicata* 5(1976), 459-466.

9. T. Brylawski, Connected matroids with the smallest Whitney numbers, *Discrete Math.* 18(1977), 243-252.

10. M. Crofton, Probability, in *Encyclopaedia Britannica*, 9th edition, 19 (1885), 768-788.

11. L. Danzer, G. Grunbaum, V. Klee, Helly's theorem and its relatives, *Proc. Symp. Pure Math.* 7, 101-180, Amer. Math. Soc. 1963.

12. J. P. Doignon, Radon partitions with k-dimensional intersection, *J. London Math. Soc.* (2), 21(1980), 365-370.

13. J. P. Doignon, Minimal numbers of circuits in affine sets, mimeographed notes: Dept. de Math., Univ. Libre de Bruxelles, 1980b.

14. J. P. Doignon, J. Reay, G. Sierksma, A Tverberg-type generalization of the Helly number of a convexity space, mimeographed notes, Dept. of Econometrics, Univ. of Groningen, 1979.

15. J. P. Doignon, G. Valette, Variations sur un Théme de Radon, dittoed notes, Univ. Libre de Bruxelles, Départment de Mathématique, 1975.

16. J. P. Doignon, G. Valette, Radon partitions and a new notion of independence in affine and projective spaces, *Mathematika* 24(1977), 86-96.

17. R. Dudley, Balls in R^d do not cut all subsets of k + 2 points, *Adv. in Math.* 31(1979), 306-308.

18. J. Eckhoff, Primitive Radon partitions, *Mathematika* 21(1974), 32-37.

19. J. Eckhoff, Radonpartitionen und konvexe Polyeder, *J. Reine Angew. Math.* 277(1975), 120-129.

20. J. Eckhoff, On a class of convex polytopes, *Israel J. Math.* 23(1976), 332-336.

21. J. Eckhoff, Radon's theorem revisited, Contributions to Geometry. *Proceedings of the Geometry Symposium in Siegen 1978.* Birkhäuser Verlag, Basel, 1979, 164-185.

22. R. Hammer, Beziehungen zwischen den Sätzen von Radon, Helly und Carathéodory bei axiomatischen Konvexitäten, *Abh. Math. Sem. Univ. Hamburg* 46(1977), 3-24.

23. W. Hare, J. Kenelly, Characterizations of Radon partitions, *Pacific J. Math.* 36(1971), 159-164.

24. W. Hare, G. Thompson, Tverberg-type theorems in convex product structures, *Geom. Metric Lin. Spaces, Proc. Conf. East Lansing 1974* (ed. L. M. Kelly). *Lecture Notes Math.* 490(1975), 212-217.

25. L. Iversland, On convex sets in R^n, (Norwegian). Cand Real. Thesis, University of Bergen, 1969.

26. R. Jamison, A development of axiomatic convexity, *Math. Tech. Report* #48, Clemson Univ.

27. R. Jamison, A theory of convexity, Doctoral Dissertation, Univ. of Washington, 1974.

28. R. Jamison, Partition numbers for trees and ordered sets, mimeographed notes, Preprint Nr. 500, Fachbereich Math., Technische Hochschule Darmstadt, 1979. To be published in *Pacific J. Math.*

29. D. Kay, E. Womble, Axiomatic convexity theory and relationships between the Carathéodory, Helly, and Radon Numbers, *Pacific J. Math.* 38(1971), 471-485.

30. V. Klee, What is the expected volume of a simplex whose vertices are chosen at random from a given convex body?, *Amer. Math. Monthly* 76(1969), 286-288.

31. D. Kramer, *Lineare Relationen zwischen Radonvektoren, primitiven Radonvektoren und f-Vektoren*, University of Dortmund, 1975.

32. D. Larman, On sets projectively equivalent to the vertices of a convex polytope, *Bull. London Math. Soc.* 4(1972), 6-12.

33. F. Levi, On Helly's theorem and the axioms of convexity, *J. Indian Math. Soc.* (N.S.) 15(1951), 65-76.

34. N. Lindquist, G. Sierksma, A generalization of the Stirling number of the second kind, mimeographed preprint WS-70/8003, Econometric Institute, Univ. of Groningen, 1980.

35. P. McMullen, Transforms, diagrams, and representations, Contributions to Geometry. *Proceedings of the Geometry Symposium in Siegen 1978.* Birkhäuser Verlag, Basel, 1979, 92-130.

36. C. Petty, Radon partitions in real linear spaces, *Pacific J. Math.* 59 (1975), 515-523.

37. J. Radon, Mengen konvexer Körper, die einen gemeinsamen Punkt enthalten, *Math. Ann.* 83(1921), 113-115.

38. J. Reay, An extension of Radon's theorem, *Illinois J. Math.* 12(1968), 184-189.

40. J. Reay, Twelve general position points always form three intersecting tetrahedra, *Discrete Math.* 28(1979), 193-199.

41. J. Reay, Several generalizations of Tverberg's theorem, *Israel J. Math.* 34, No. 3 (1979), 238-244.

42. G. Shephard, Neighbourliness and Radon's theorem, *Mathematika* 16(1969), 273-275.

43. G. Sierksma, Axiomatic convexity theory and the convex product space, Doctoral Dissertation, Univ. of Groningen, 1976.

44. G. Sierksma, Relationships between Carathéodory, Helly, Radon and exchange numbers of convexity spaces, *Nieuw Arch. Wisk.* (3) 25(1977), 115-132.

45. G. Sierksma, Convexity without linearity; the Dutch cheese problem, mimeographed notes, Dept. of Econometrics, Univ. of Groningen, 1979.

46. G. Sierksma, Generalized Radon partitions in convexity spaces, mimeographed Preprint WS-72/8005, Econometrics Institute, Univ. of Groningen, 1980.

47.   J. Strangeland, Convex properties of finite sequences of points in R^n, (Norwegian). Cand. Real. Thesis, University of Bergen, 1978.

48.   H. Tverberg, A generalization of Radon's theorem, *J. London Math. Soc.* 41(1966), 123-128.

49.   H. Tverberg, A further generalization of Radon's theorem, *J. London Math. Soc.* 43(1968), 352-354.

50.   R. Winder, Partitions of n-space by hyperplanes, *SIAM J. Appl. Math.* 14(1966), 811-818.

GENERALIZATIONS OF HELLY'S THEOREM; OPEN PROBLEMS

Gerard Sierksma

Subdepartment of Mathematics
Econometric Institute
University of Groningen
Groningen, The Netherlands

1. INTRODUCTION

Helly's theorem was first published by Johann Radon in 1921. Eduard Helly published his own proof in 1925, thirteen years after he discovered it. The theorem asserts that each family of convex sets in R^d, which is finite or whose members are compact, has a nonempty intersection, provided each subfamily of at most $d + 1$ sets has nonempty intersection.

This formulation of Helly's theorem can be found in the famous paper of L. Danzer, B. Grünbaum, and V. Klee [3], called "Helly's Theorem and its Relatives" (1963); the paper contains an extensive bibliography. Restricting Helly's theorem to *finite* families of convex sets, it is clear that the theorem is then formulated completely in terms of convex sets, their intersections, and the dimension d of the underlying space. This suggests defining the Helly number of an *aligned space*, which is a pair (X,C) with X a set and C a collection of subsets of X, called *convex sets*, such that

1. ϕ, $X \in C$;

2. C is closed under intersections;

3. C is closed under nested unions (a nest is a family of sets totally ordered by inclusion).

C is then called an *alignment* on the set X. The *C-hull* of a set S in X is defined as $C(S) = \cap\{A \in C \mid S \subset A\}$. The axioms 1 and 2 are first used by F. W. Levi [18] in 1951, and later on by many other authors, e.g., Eckhoff [8], Jamison [12], Kay and Womble [16], and Sierksma [23]. The concept of alignment is introduced by Jamison [12]. Hammer [10] has shown that axiom 3 is equivalent to the "*domain finiteness* condition" which says that for each S in X, $C(S) = \cup\{C(T) \mid T \subset S, |T| < \infty\}$ ($|T|$ is the cardinality of T). Instead of the term alignment we find in the literature the terms "algebraic closure system" [2] and "domain-finite convexity space" [4, 7, 8, 16, 22-25].

The *Helly number* h of an aligned space (X,C) is usually defined to be the infimum of all nonnegative integers k so that:

a. The intersection of any finite collection of convex sets is nonempty provided the intersection of each subcollection of it consisting of at most k elements is nonempty.

Note that the Helly number of R^d with the ordinary alignment is equal to $d + 1$. It was shown by Berge and Duchet [1] and Sierksma [22] that h may equivalently be defined to be the infimum of all nonnegative integers k so that:

b. Each S in X with $|S| \geq k + 1$ has the property that $\cap\{C(S \setminus \{x\}) \mid x \in S\} \neq \phi$

Instead of a finite collection in (a) one can take a collection of $k + 1$ sets, and in (b) it suffices to take $|S| = k + 1$; see Sierksma [23], Theorem 8.1. Note that, in general, $0 \leq h \leq \infty$ and that h = 0 if and only if $X = \phi$. If (X,C) is a T_1 space, i.e., each single element of X is convex, then h = 1 if and only if $|X| = 1$.

Closely related to Helly's theorem is the classical theorem of Radon; see e.g., Danzer, Grünbaum, Klee [3]. The *Radon number* r of an aligned space (X,C) is the infimum of all positive integers k with the property that:

Each set S in X with $|S| \geq k$ admits a partition $S = S_1 \cup S_2$ with $S_1 \cap S_2 = \phi$ and such that $C(S_1) \cap C(S_2) \neq \phi$.

Such a partition is called a Radon partition of S. Radon's theorem says that $r = d + 1$. The relationship between the Helly and Radon numbers is expressed by Levi's Theorem, namely that $h \leq r - 1$; see e.g., [24].

2. A TVERBERG-TYPE GENERALIZATION OF THE HELLY NUMBER

Radon's theorem was generalized in 1966 by H. Tverberg [26]; instead of 2-partitions he has investigated arbitrary τ-partitions. Similarly, the Radon number can be generalized as follows (see [7]): For $\tau \geqslant 1$ the τ-*Radon number* $r(\tau)$ of an aligned space $(X, \mathcal{C})$ is the infimum of all positive integers k with the property that: Each set S in X with $|S| \geqslant k$ admits a τ-partition $S = S_1 \cup \cdots \cup S_\tau$ into pairwise disjoint sets S_i such that $\mathcal{C}(S_1) \cap \cdots \cap \mathcal{C}(S_\tau)$ $\neq \phi$. Such a τ-partition of S is called a *Radon τ-partition* of S. Tverberg's theorem states that $r(\tau) = (\tau - 1)(d + 1) + 1$ for the ordinary aligned space (R^d, conv). To give a similar generalization of the Helly number we first introduce the σ-core of a set S in X ($\sigma \geqslant 0$), namely

$$\text{core}_\sigma (S) = \cap \{ \mathcal{C}(S \setminus M) \mid M \subset S, \ |M| \leqslant \sigma \}$$

(See Doignon, Reay, Sierksma [7] and Jamison [13].) Note that $\mathcal{C}(S) = \text{core}_0 (S)$ $= \text{core}_1 (S) \supset \cdots$. Helly's theorem says that $\text{core}_1 (S) \neq \phi$ provided $|S| \geqslant d + 2$ for $S \subset R^d$. The σ-*Helly number* $h(\sigma)$ of an aligned space $(X, \mathcal{C})$ is the infimum of all nonnegative integers k such that

(b')   For each S in X with $|S| \geqslant k + 1$, $\text{core}_\sigma (S) \neq \phi$

Thus $h(\sigma) = \infty$ if there are arbitrarily large subsets of X with empty σ-core. Note that $h(1) = h$, and that $h(\sigma) = 0$ iff $X = \phi$ or $\sigma = 0$.

PROBLEM 1. Find a formulation of the σ-Helly number equivalent to (b') in terms of intersections of collections of convex sets, similar to (a).

The following theorem generalizes Levi's Theorem; a proof can be found in [7].

THEOREM 2.1. In any aligned space $(X, \mathcal{C})$ the following holds:

$$h(\sigma) \leqslant r(\sigma + 1) - 1$$

A result similar to Tverberg's theorem is the following; also see [7].

THEOREM 2.2 For any ordinary aligned space of dimension d it follows that

$$h(\sigma) = (d + 1)\sigma$$

The next result, which holds *mutatis mutandis* for the τ-Radon number, shows the monotonicity of the function h; see [7].

THEOREM 2.3. For any aligned space with σ-Helly number $h(\sigma) \leqslant |X| - 1$ the following holds:

$$h(\sigma) \leqslant h(\sigma + 1) - 1$$

In Eckhoff [8] the following conjecture on the τ-Radon number $r(\tau)$ in terms of the 2-Radon number r can be found:

$$r(\tau) \leqslant (\tau - 1)(r - 1) + 1$$

The conjecture is strongly supported by many examples; see e.g., Jamison [13]. For the σ-Helly number we have the following result; see [7].

THEOREM 2.4. For any aligned space $(X, \mathcal{C})$ and $\sigma \geqslant 1$ it follows that

$$\min\{|X|, h + \sigma - 1\} \leqslant h(\sigma) \leqslant \sigma h$$

In the above theorem h is the 1-Helly number. It can be shown that both bounds for $h(\sigma)$ are sharp; see [7].

One of the most famous open problems in axiomatic convexity is the "sharpness" of the Eckhoff and Jamison inequality, namely $r \leqslant c(h - 1) + 2$, where c is the Carathéodory number. For a proof of this inequality see Sierksma [23]. For the τ-Radon number one can show that $r(\tau) \leqslant (\tau - 1)ch - c + 2$; see [7].

PROBLEM 2. Find a relationship between $r(\tau)$, $h(\tau - 1)$ and c, similar to the Eckhoff and Jamison inequality. Conjecture: $r(\tau) \leqslant ch(\tau - 1) - c + 2$.

The sharpness of the Eckhoff and Jamison theorem has stimulated the investigation of the product and sum spaces; see e.g., [8, 22, and 25]. Let $(X_1, \mathcal{C}_1)$ and $(X_2, \mathcal{C}_2)$ be aligned spaces; their *aligned product space* is the pair $(X_1 \times X_2, \mathcal{C}_1 \oplus \mathcal{C}_2)$ where $\mathcal{C}_1 \oplus \mathcal{C}_2 = \{A \times B \mid A \in \mathcal{C}_1, B \in \mathcal{C}_2\}$.

THEOREM 2.5. Let (X_1, C_1) and (X_2, C_2) be nonempty aligned spaces with σ-Helly numbers $h_1(\sigma)$ and $h_2(\sigma)$, respectively, and assume that $h_i(\sigma) + 1 \leqslant |X_i|$ for $i = 1, 2$. Then the σ-Helly number $h(\sigma)$ of $(X_1 \times X_2,\ C_1 \oplus C_2)$ satisfies

$$\max\{h_1(\sigma), h_2(\sigma)\} \leqslant h(\sigma) \leqslant \sigma \max\{h_1(\sigma), h_2(\sigma)\} - \sigma^2 + \sigma$$

Note that for $\sigma = 1$ we have $h = \max\{h_1, h_2\}$, which is Theorem 8.3 of [23]. A proof of the above theorem can be found in [7].

The concept of aligned sum space is introduced in Sierksma [25]. Let (X_1, C_1) and (X_2, C_2) be aligned spaces; their *aligned sum space* is the pair $(X_1 \cup X_2,\ C_1 + C_2)$ where $C_1 + C_2 = \{(A \setminus X_2) \cup (B \setminus X_1) \cup (A \cap B) \mid A \in C_1,\ B \in C_2\}$. When $X_1 = X_2$ we have $C_1 + C_2 = \{A \cap B \mid A \in C_1,\ B \in C_2\}$ which is the join alignment on X_1 $(= X_2)$. If $X_1 \cap X_2 = \phi$ we have $C_1 + C_2 = \{A \cup B \mid A \in C_1,\ B \in C_2\}$.

THEOREM 2.6. Let (X_1, C_1) and (X_2, C_2) be aligned spaces with $X_1 \cap X_2 = \phi$ and let $h_1(\sigma)$ and $h_2(\sigma)$ be the respective σ-Helly numbers. Then the σ-Helly number $h(\sigma)$ of $(X_1 \cup X_2,\ C_1 + C_2)$ satisfies

$$h(\sigma) = h_1(\sigma) + h_2(\sigma)$$

With the help of the aligned product and sum spaces it can be shown that in case $h \leqslant e - 1$ the inequality

$$c \leqslant \max\{h, e - 1\}$$

is sharp where e is the exchange number; (see [22 and 25]). The sharpness in case $h > e - 1$ is still open. If $e \leqslant c$ it can be shown that (see Sierksma [25]):

$$r \leqslant (c - 1)(h - 1) + 3$$

Combining aligned products and sums it is shown in [25] that this inequality is sharp for $c < 4$. For $c \geqslant 4$ the sharpness is still questionable. Clearly, the notions of aligned product and aligned sum space can be extended to products and sums of more than two spaces with similar results for the σ-Helly numbers.

PROBLEM 3. Given the σ-Helly numbers for the aligned spaces (X_1, C_1) and (X_2, C_2), what is the σ-Helly number for the aligned sum space $(X_1 \cup X_2, C_1 + C_2)$, without the restriction $X_1 \cap X_2 = \phi$? This problem is also open for the Carathéodory, Radon, and Exchange numbers. Some partial results can be found in Degreef [4].

PROBLEM 4. Let $h_1(\sigma)$ and $h_2(\sigma)$ be the σ-Helly numbers of (X, C_1) and (X, C_2), respectively. If $C_1 \subset C_2$, what is the relationship between $h_1(\sigma)$ and $h_2(\sigma)$? This problem is also open for the other numbers.

In Kay and Womble [16] the concept of C-halfspaces is introduced: a set H in X is called a *C-halfspace* iff $H \in C$ and $X \setminus H \in C$. The aligned space satisfies the *separation property* iff for each two disjoint convex sets A and B there exists a C-halfspace H such that $A \subset H$ and $B \subset X \setminus H$. Kay and Womble then proved the following characterization of h: In any T_1 aligned space having the separation property, the following two conditions are equivalent:

    (i) (X, C) has Helly number $h \leq k$;

    (ii) If $S \subset X$ with $|S| = k + 1 \geq 3$, then there exists an element $p \in X$ such that every C-halfspace containing at least k elements of S also contains p.

PROBLEM 5. Find a similar characteriztion of $h(\sigma)$. Conjecture: (ii)' $h(\sigma) \leq k$ iff for $S \subset X$ with $|S| = k + 1 \geq 3$ there exists $M \subset X$ with $|M| = \sigma$ such that every C-halfspace containing at least k elements of S also contains M.*

We conclude this section by giving the σ-Helly numbers of several aligned spaces (with $|X| = \infty$):

    1. $(X, C) = (R^d, \text{conv})$: $h(\sigma) = (d + 1)\sigma$

    2. $C = \{\phi, X\}$: $h(\sigma) = \sigma$

    3. $C = \{A \mid A \subset X\}$: $h(\sigma) = \infty$

    4. $C = \{X\} \cup \{A \mid A \subset X, |A| \leq k\}$ with $k \geq 2$: $h(\sigma) = k + \sigma$

---

*Recently J. P. Doignon has shown that the conjecture is false and that the correct characterization is:

(ii)' For $S \subset X$ with $|S| \geq k + 1$ there exists an element $p \in X$ such that every C-halfspace containing at least $k + 1 - \sigma$ points of S also contains p.

5. $X = R^d$, C all closed sets in R^d: $h(\sigma) = \infty$

6. $X = R^d$, C all closed convex sets in R^d: $h(\sigma) = (d + 1)\sigma$

7. $M \neq \phi$, $C = \{A \mid M \subset A\}$: $h(\sigma) = \sigma$

8. $|M| = m \geqslant 2$, $C = \{X\} \cup \{A \mid M \not\subset A\}$: $h(\sigma) = \max\{m,\sigma\}$

An interesting aligned space occurs by taking the restriction of the ordinary alignment to the Gaussian integers (points in R^d with integer co-efficients), introduced by J. P. Doignon [5]. He shows that $h = 2^d$. Also see Jamison-Waldner [13] for a different proof.

PROBLEM 6.   What is the σ-Helly number of the aligned space (R^d,C) where C is the restriction of the ordinary alignment to the Gaussian integers of R^d. Is $h(\sigma) = 2^d\sigma$?*

3. ON THE DIMENSION OF THE σ-CORE

In [20, 21] J. R. Reay has introduced the concept of (τ,k)-partition: Any set (family of elements) in R^d has a (τ,k)-*partition* $S = S_1 \cup \cdots \cup S_\tau$ if the sets S_i are pairwise disjoint and $\dim[\cap_{i=1}^\tau \mathrm{conv}(S_i)] \geqslant k$. An element of $\cap_{i=1}^\tau \mathrm{conv}(S_i)$ is called a (τ,k)-*divisible* point of S. Also see J. P. Doignon [6].

FIGURE 1

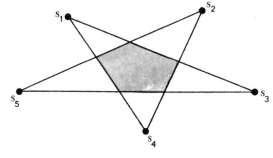

By $D_\tau(S)$ Reay denotes the set of elements p in X such that $p \in \cap_{i=1}^\tau C(S_i)$ for *some* Radon τ-partitions $S = S_1 \cup \cdots \cup S_\tau$ of S. Note that $C(S) = D_1(S) \supset D_2(S) \supset \cdots$. According to Reay, the dimension of $D_\tau(S)$ in R^d is the maximum of the dimensions of the intersections of the several Radon τ-partitions of S. For the five points in Fig. 1, $\dim D_2(S) = 1$ ($S = \{s_1,s_2,s_3,s_4,s_5\}$).

---

*Recently J. P. Doignon has shown that the conjecture is true, by taking the 2^d vertices of a fundamental cube in Z^d and $\sigma - 1$ more lattice points on the 2^d "symmetric rays" starting in the vertices of the cube; these $\sigma2^d$ points have nonempty σ-core.

Note that dim core$_1$(S) = 2, and that $D_2(S) \subsetneq \text{core}_1(S)$.

THEOREM 3.1. For any aligned space (X,C) and S in X the following holds:

$$C(D_\tau(S)) \subset \text{core}_{\tau-1}(S)$$

As we have seen above, the converse of this theorem is not true in general. The proof of 3.1 is a direct consequence of 2.1. Using an argument similar to Birch's, Reay [21] has shown the following theorem.

THEOREM 3.2. If (X,C) = (R^2,conv) and S $\subset$ R^2 then

$$\text{conv } D_\tau(S) = \text{core}_{\tau-1}(S)$$

As is customary, the dimension of a subset in R^d is taken as the dimension of its affine hull. Three additional problems on the σ-core follow.

PROBLEM 7. Find a convexity space (X,C) such that for some $n_0 \geqslant 0$ and each $n \geqslant n_0$ there is a set S in X with $|S| = n$ and $C(D_\tau(S)) \neq \text{core}_{\tau-1}(S)$.

PROBLEM 8. Is conv $D_\tau(S) = \text{core}_{\tau-1}(S)$ for each S in R^d with $|S| = (\tau - 1)$ $(d + 1) + 1$? Note that if this is the case, we have an alternative proof of Tverberg's Theorem.

PROBLEM 9. Is the number of Radon τ-partitions dependent on the number of vertices of core$_{\tau-1}$(S) for finite sets S in R^d? I.e., for the σ-core being a single point does the set have the minimum number of Radon τ-partitions?

THEOREM 3.3. For each S in R^d with $|S| \geqslant 2\sigma d + 2$,

$$\text{dim core}_\sigma(S) \geqslant 1$$

The proof follows directly from Reay [21] Theorem 5 together with 3.2. This theorem also says that each set in R^d with at least $2\sigma d + 2$ elements admits at least two elements in the σ-core. That $2\sigma d + 2$ is the best possible result follows from the fact that $2\sigma d + 1$ points with σ points on each of the positive as well as on the negative axes together with the origin have as σ-core the origin, so its σ-core is 0-dimensional.

In the following theorem we answer the question: What is the minimum number of points with a d-dimensional σ-core. H^+ and H^- are the open half-spaces determined by the hyperplane H.

THEOREM 3.4. Let $S \subset R^d$. Then (i) ⇒ (ii) ⇒ (iii), with:

(i)   dim $core_\sigma(S) = d$;

(ii) For each hyperplane H in R^d, $|H^+ \cap S| \geq \sigma + 1$, or $|H^- \cap S| \geq \sigma + 1$, or both;

(iii) $|S| \geq 2\sigma + d + 1$.

Proof. Take any S in R^d.

(i) ⇒ (ii): Take any hyperplane H in R^d with $|H^+ \cap S| \leq \sigma$ and $|H^- \cap S| \leq \sigma$. Then, clearly, $core_\sigma(S) \subset H$, and therefore, dim $core_\sigma(S) \leq$ dim H = d - 1, which is a contradiction. Hence, each hyperplane has at least σ + 1 points of S on at least one (open) side.

(ii) ⇒ (iii): If S has at most 2σ + d elements, there is a hyperplane H spanned by at least d elements of S and with at most σ elements in each open halfspace H^+ and H^-. This contradicts (ii). Therefore $|S| \geq 2\sigma + d + 1$. □

An element p in X is called an *extreme* element of S iff p $\notin$ conv(S \ p). Also see the next section for a more extensive use of extreme elements and sets in relation to the σ-core.

THEOREM 3.5. Let $S \subset R^2$ with $|S| \geq 2\sigma + 3$ and with at least 3σ + 1 extreme points. Then dim $core_\sigma(S) = 2$.

The proof of this theorem is omitted because it is similar to the proof of the next theorem.

The example in Fig. 2 shows that for σ = 2 we need in fact more than 6 *extreme* points in the above theorem; note that the 2-core is the point "7."

The independence condition in Theorem 3.6(ii) plays a part in the next theorem. In the proof is used Peterson's method of projections; see Peterson [19] and also Doignon [6]. Note that the smallest sets for which (ii) holds with σ = 0 are *simplices*.

THEOREM 3.6. Let $S \subset R^3$ with $|S| \geq 4\sigma + 3$ and with at least 4σ + 1 extreme points; moreover, suppose that for each hyperplane H, or $|H^+ \cap S| \geq \sigma + 1$, or $|H^- \cap S| \geq \sigma + 1$, or both are. Then, dim $core_\sigma(S) = 3$.

FIGURE 2

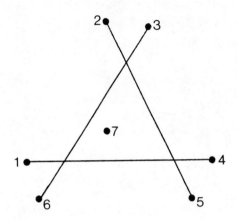

Proof. Take any sets T_1, T_2, T_3, T_4 in S with $|T_i| \leqslant \sigma$ for $i = 1,2,3,4$. Then $|\cup_{i=1}^{4} T_i| \leqslant 4\sigma$, so there is an extreme point p in S with $p \notin \cup_{i=1}^{4} T_i$. Let H be a hyperplane with $p \in H^-$ and $conv(S \setminus p) \subset H^+$. Then, define S* = $H \cap \{[p,s] \mid s \in S \setminus p\}$, hence, $|S*| \geqslant 4\sigma + 2$. Theorem 3.3 then implies that dim $core_\sigma(S*) \geqslant 1$. Clearly, the dimension of the σ-core of S* cannot be e-qual to 1, because, otherwise, there would exist a line L with $core_\sigma S* \subset L$ in H with less than $\sigma + 1$ points on both sides, and then the hyperplane $aff(p \cup L)$ has less than $\sigma + 1$ on both sides, which contradicts the assump-tion of the theorem. Therefore, dim $core_\sigma(S*) = 2$, and so we have $\{p\} \cup$ $core_\sigma(S*) \subset \cap_{i=1}^{4} conv(S \setminus T_i)$. Grünbaum [11] then implies that dim $core_\sigma(S)$ = 3. □

The above theorem makes use of the first part of Grünbaum's theorem in the paper "The Dimension of Intersections of Convex Sets" (1962). Katchalski (1971) has pointed out in his paper with the same title that the second part of Grünbaum's theorem is wrong; see [14 and 15]. The following theorem is Katchalski's generalization of Helly's theorem. The number $\alpha(k,d)$ is defined as follows:

$$\alpha(k,d) = d + 1 \quad \text{if } k = 0$$

$$\alpha(k,d) = \max\{d + 1, 2d - 2k + 2\} \quad \text{if } 1 \leqslant k \leqslant d$$

THEOREM 3.7. If G is any finite family of convex subsets of R^d, and for every subfamily F with $|F| \leqslant \alpha(k,d)$, dim $\cap F \geqslant k$, then dim $\cap G \geqslant k$.

Let $0 \leqslant k \leqslant d$. A set S in R^d with $|S| \geqslant k + 1$ is called k-*independent* (see e.g., [21]) iff for each T in S with $|T| = k + 1$, dim aff(T) = k. Note that each nonempty set is 0-independent and that "d-independent" is equivalent to "*in general position.*"

THEOREM 3.8. Let $0 \leqslant k \leqslant d$ and $S \subset R^d$ with $|S| \geqslant \sigma\alpha(k,d) + k + 1$ and S k-independent. Then

$$\dim \mathrm{core}_\sigma (S) \geqslant k$$

Proof. Take any α (= $\alpha(k,d)$) sets $T_1.,,,.T_\alpha$ in S with $|T_i| \leqslant \sigma$ for each i = 1,...,n. As $|S| \geqslant \sigma\alpha(k,d) + k + 1$, and $|\cup_{i=1}^\alpha T_i| \leqslant \sigma\alpha(k,d)$, it follows that $|S \setminus \cup_{i=1}^\alpha T_i| \geqslant k + 1$.

As S is k-independent, it follows that dim aff($S \setminus \cup_{i=1}^\alpha T_i) \geqslant k$. Clearly, $S \setminus \cup_{i=1}^\alpha T_i \subset \cap_{i=1}^\alpha (S \setminus T_i) \subset \cap_{i=1}^\alpha \mathrm{conv}(S \setminus T_i)$. Hence, dim $\cap_{i=1}^\alpha$ conv $(S \setminus T_i) \geqslant \dim(S \setminus \cup_{i=1}^\alpha T_i) \geqslant k$. Katchalski's theorem then implies that dim core$_\sigma$(S) = dim $\cap\{\mathrm{conv}(S) \setminus T) \mid T \subset S, |T| \leqslant \sigma\} \geqslant k$. $\square$

Note that for k = 0 in the above theorem the condition "S is k-independent" can be deleted; the theorem then reduces to Theorem 2.2. For k = 1 the theorem also holds without "k-independent"; see Theorem 3.3. In the following corollary k = d; also see Reay [21], Theorem 4.

COROLLARY 3.9. Let $S \subset R^d$ with $|S| \geqslant (\sigma + 1)(d + 1)$ and S in general position. Then

$$\dim \mathrm{core}_\sigma (S) = d$$

The conditions on the set S in Theorem 3.8 are not necessary. Consider for instance, the seven points in the plane as shown in Fig. 3. It is clear that the dimension of the 2-core of these seven points is 2, although there are just seven points. Note that the points are in general position.

For the case $\sigma = 1$ several more theorems can be obtained by taking into account Theorem 3.1 and making use of the theorems on (2,k)-divisibility in Doignon [6] and Reay [20, 21]. On the other hand, it is remarkable that the theorems on the σ-Helly number and on the dimension of the σ-core are generally not very difficult to prove, whereas the similar assertions on the τ-Radon

FIGURE 3

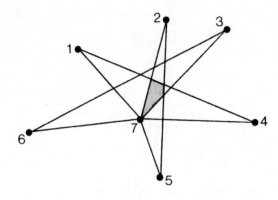

(Eckhoff's conjecture) and on the dimension of (τ,k)-divisible points are to
a great extent still open, although there is a quite strong relationship
between the Helly and Radon relatives. Is the main reason for this gap in
difficulty the fact that the σ-core is convex and the set of (τ,k)-divisible
points are *not* convex in general?

PROBLEM 10. Find necessary and sufficient conditions for the σ-core of a
set S in R^d to be k-dimensional $(0 \leq k \leq d)$.

4. REAY- AND LARMAN-TYPE GENERALIZATIONS OF HELLY'S THEOREM

In Reay [21] a generalization of Radon's and Tverberg's theorem is given
in the ordinary aligned space (R^d, conv). Along this line we give the following generalization of the Radon number.

Suppose $n,\tau \geq 1$ are integers. The (τ,n)-*Radon number* $r(\tau,n)$ of an a-
ligned space (X,C) is for $n \leq \tau$ the infimum of all positive integers k such
that for each $S \subset X$ with $|S| \geq k$ there is a τ-partition $S = S_1 \cup \cdots \cup S_\tau$
of S into pairwise disjoint sets with the property that

$$C(S_{i_1}) \cap \cdots \cap C(S_{i_n}) \neq \phi$$

for each $i_1,\ldots,i_n \in \{1,\ldots,\tau\}$; for $n > \tau$ we define $r(\tau,n) = r(\tau,\tau)$. Note
that $r(\tau,\tau)$ is the τ-Radon number (Tverberg's generalization), and that

$r(\tau,1) = \tau$.

For a generalization of Levi's theorem 2.1 similar to Reay's generalization of the Radon and Tverberg theorems, we introduce for $n, \sigma \geqslant 0$ the (σ,n)-*Helly number* $h(\sigma,n)$ of an aligned space (X,C) as the infimum of all nonnegative integers k such that each $S \subset X$ with $|S| \geqslant k + 1$ has the property that

$$C(S \setminus M_1) \cap \cdots \cap C(S \setminus M_n) \neq \phi$$

for each $M_1,\ldots,M_n$ in S with $|M_i| \leqslant \sigma$ and each $i = 1,\ldots,n$. If $|S| = p \geqslant \sigma$ and $\binom{p}{\sigma} = \sigma'$, then $h(\sigma,\sigma')$ is the σ-Helly number. Furthermore, $h(\sigma,1) = \sigma$. Levi's theorem 2.1 can be generalized as follows.

THEOREM 4.1. For any aligned space (X,C) the following holds:

  1. $h(\sigma,n) \leqslant \min\{n\sigma, h(\sigma)\}$

  2. $h(\sigma,n) = h(\sigma) \leqslant \sigma h(1)$ if $n \geqslant h(1)$

  3. $h(\sigma,n) \leqslant n\sigma < \sigma h(1)$ if $n < h(1)$

Proof. (1) Take any $S \subset X$ with $|S| \geqslant 1 + \min\{n\sigma, h(\sigma)\}$. If $|S| \geqslant 1 + h(\sigma)$, then $\phi \neq \mathrm{core}_\sigma(S) \subset \cap\{C(S \setminus M_{i_k}) \mid M_{i_k} \subset S, |M_{i_k}| \leqslant \sigma$ for $k = 1,\ldots,n\}$ and hence $h(\sigma,n) \leqslant h(\sigma)$. If $|S| \geqslant 1 + n\sigma$ then, clearly, for each n sets $M_1,\ldots,M_n$ in S with $|M_i| \leqslant \sigma$ there is an element $p \in S$ such that $p \in S \setminus (M_1 \cup \cdots \cup M_n) = (S \setminus M_1) \cap \cdots \cap (S \setminus M_n)$, so that $\cap\{C(S \setminus M_i) \mid i = 1,\ldots,n\} \neq \phi$ and it follows that $h(\sigma,n) \leqslant n\sigma$. Hence, $h(\sigma,n) \leqslant \min\{n\sigma, h(\sigma)\}$. (2) Consider the family $F = \{C(S \setminus M) \mid M \subset S, |M| \leqslant \sigma\}$. As $n \geqslant h(1)$ it follows that for each $F_0 \subset F$ with $|F_0| = h(1)$, $\cap F_0 \neq \phi$. Helly's theorem then implies $\cap F \neq \phi$. Hence, $h(\sigma) \leqslant h(\sigma,n)$. Combining this with (1) and Theorem 2.4 yields the desired result. (3) For $n < h(1)$ it follows from (1) that $h(\sigma,n) \leqslant n\sigma < \sigma h(1)$. $\square$

The following theorem is a generalization of Levi's theorem for the (σ,n)-Helly and (τ,n)-Radon numbers.

THEOREM 4.2 In any aligned space (X,C) the following holds:

$$h(\sigma,n) \leqslant r(\sigma + 1, n) = 1$$

Proof. Equality holds in the trivial case $h(0,n) = r(1,n) - 1 = 0$, and $h(\sigma,n) = |X| = r(\sigma+1, n) - 1$ when $\sigma \geq |X|$. If a finite space X has no Radon τ-partition (so that $r(\tau) = |X| + 1$) then the theorem again is trivial, since $h(\sigma,n) \leq |X|$ for any $\sigma \geq 0$. Thus we may assume $\sigma \geq 1$ and $r(\sigma+1, n)$ is finite and is at most $|X|$ in case X is a finite set. First let $n \leq \sigma + 1$. Take any $S \subset X$ with $|S| = r(\sigma+1, n)$ and let $M_1,\ldots,M_n$ be n subsets of S with $|M_i| \leq \sigma$ for $i = 1,\ldots,n$. Furthermore, let $S = S_1 \cup \cdots \cup S_{\sigma+1}$ be a $(\sigma + 1)$-partition of S with the property that for each n of its sets the convex hulls have non-empty intersection, i.e., $C(S_{i_1}) \cap \cdots \cap C(S_{i_n}) \neq \phi$ for each $i_1,\ldots,i_n \in \{1, 2,\ldots,\sigma + 1\}$. For each M_i there is a set S_{j_i} with $j_i \in \{1,\ldots,\sigma + 1\}$ and such that $S_{j_i} \subset S \setminus M_i$. Hence, $C(S_{j_1}) \subset C(S \setminus M_1), \ldots, C(S_{j_n}) \subset C(S \setminus M_n)$, and therefore we have $\cap_{i=1}^n C(S_{j_i}) \subset \cap_{i=1}^n C(S \setminus M_i)$. As $\cap_{i=1}^n C(S_{j_i}) \neq \phi$, it follows that $\cap_{i=1}^n C(S \setminus M_i) \neq \phi$. This implies that in fact $h(\sigma,n) \leq r(\sigma+1, n) - 1$. For $n > \sigma + 1$, it follows from 4.1 that $h(\sigma,n) \leq h(\sigma) \leq r(\sigma+1) - 1 = r(\sigma+1, \sigma+1) - 1 = r(\sigma+1, n) - 1$. $\square$

THEOREM 4.3. For the ordinary convexity space with dimension d the following holds:

$$h(\sigma,n) = \min\{n\sigma, (d + 1)\sigma\}$$

Proof. If $n \geq d + 1 = h(1)$, then 4.1 (2) implies $h(\sigma,n) = (d + 1)\sigma$. Now let $n < d + 1$. Take some simplex in R^d and let S be the set consisting of n vertices of a simplex with multiplicity σ; let $M_1,\ldots,M_n$ be the vertices with multiplicities σ (S is a family of elements). Then, clearly, $\text{conv}(S \setminus M_1) \cap \cdots \cap \text{conv}(S \setminus M_n) = \phi$, and therefore $h(\sigma,n) \geq n\sigma$. As $n\sigma < (d + 1)\sigma$ it follows from (1) that $h(\sigma,n) = n\sigma$. Hence, $h(\sigma,n) = \min\{n\sigma,(d + 1)\sigma\}$. $\square$

Let $S \subset X$ with $|S| = m \geq \sigma$. The σ-core of S is then the intersection of the $\binom{m}{\sigma}$ sets $C(S \setminus M)$ with $M \subset S$ and $|M| = \sigma$. In general one can take fewer than $\binom{m}{\sigma}$ sets. By n* we shall mean the minimum number of sets M_i in S with $|M_i| = \sigma$ such that $\text{core}_\sigma(S) = C(S \setminus M_1) \cap \cdots \cap C(S \setminus M_{n*})$.

We now pay some attention to the nature of the sets M_i that are needed to construct the σ-core. To that end we introduce the following concepts.

Let S be a subset of X. A subset T of S is called an *extreme set* of S if

$$T \cap C(S \setminus T) = \phi$$

An element p of S is called an *extreme element (point)* of S iff {p} is an extreme set of S. The set T is called an *extreme cluster* of S iff

$$C(T) \cap C(S \setminus T) = \phi$$

In the following figure $S = \{s_1, \ldots, s_6\} \cup S_7$ with $|S_7| = 5$. The set S_7 is an extreme set as well as an extreme cluster in S. The set $\{s_2, s_6\}$ is extreme but not an extreme cluster. The element s_3 is not extreme.

FIGURE 4

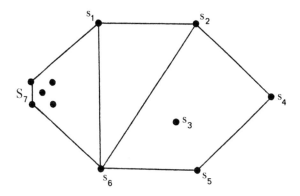

On the other hand consider the aligned space (X, C) with $C = \{\phi\} \cup \{A \mid A \subset X, a \in A\}$ for some fixed element $a \in X$. Then for each $S \subset X$ with $|S| > \sigma$ and for each $\sigma \geq 1$ we have $\mathrm{core}_\sigma(S) = \{a\}$. If $a \notin S$ then each subset of S is extreme; if $a \in S$ then the only subset of S that is not extreme is $\{a\}$. Note that there are no extreme clusters.

Note that an "extreme cluster" is always "extreme." Also note that both components of a Radon 2-partition of a set are extreme clusters, i.e., T is an extreme cluster of S iff $S \setminus T$ is an extreme cluster of S. By $E_\sigma(S)$ we denote the collection of all extreme sets in S with at most σ elements.

THEOREM 4.4. Let (X, C) be an aligned space and $S \subset X$ with $|S| \geq \sigma$. Then the following holds:

$$\mathrm{core}_\sigma(S) = \cap \{C(S \setminus E) \mid E \in E_\sigma(S)\}$$

Proof. Take any S in X with $|S| \geq \sigma$, and let M be a subset of S with

$|M| \leqslant \sigma$. Define $M = M_1 \cup E_1$ with $M_1 \subset C(S \setminus M)$ and $E_1 \cap C(S \setminus M) = \phi$. Clearly, $E_1 \in E_\sigma(S)$, and $C(S \setminus M) = C(S \setminus (M_1 \cup E_1)) = C(S \setminus M_1) \setminus E_1) \subset C(S \setminus E_1)$. Hence, $\mathrm{core}_\sigma(S) = \cap\{C(S \setminus M) \mid M \subset S, |M| \leqslant \sigma\} \subset \cap\{C(S \setminus E) \mid E \in E_\sigma(S)\}$.

As the converse of this inclusion holds trivially, we have that in fact $\mathrm{core}_\sigma(S) = \cap\{C(S \setminus E) \mid E \in E_\sigma(S)\}$. $\square$

The σ-core of infinite sets in R^d is somewhat different from the σ-core of a finite set in R^d. Consider for instance the 1-core of the closed convex circle disk C and the closed square D in R^2. The 1-core of C or D is equal to the σ-core of C or D for each $\sigma \geqslant 1$. Note that the σ-core of C is the open circle disk and that the σ-core of D is D minus the four vertices. In any case, the σ-core is the set itself minus its extreme points. The following theorem characterizes the σ-core of those "infinite" sets.

FIGURE 5

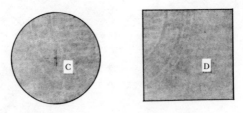

THEOREM 4.5. Let (X,C) be an aligned space and $S \subset X$ with $|S| \geqslant \sigma$. Then the following assertions are equivalent:

    (i) $C(S) \setminus E = C(S \setminus E)$ for each $E \in E_\sigma(S)$;

    (ii) $\mathrm{core}_\sigma(S) = C(S) \setminus \cup E_\sigma(S)$.

Proof. (i) $\Rightarrow$ (ii): $C(S) \setminus \cup E_\sigma(S) = \cap\{C(S) \setminus E \mid E \in E_\sigma(S)\} = \cap\{C(S \setminus E) \mid E \in E_\sigma(S)\} = \mathrm{core}_\sigma(S)$ (according to 4.4).

(ii) $\Rightarrow$ (i): As $C(S \setminus E) \cup E \subset C(S)$ for each $E \subset S$, it suffices to show that $C(S) \setminus E \subset C(S \setminus E)$. Take any $p \in C(S) \setminus E$. Then, $p \in C(S) = [\mathrm{core}_\sigma(S)] \cup [\cup E_\sigma(S)]$. If $p \in \cup E_\sigma(S) \subset S$, then $p \in S \setminus E \subset C(S \setminus E)$, and we are done. If $p \in \mathrm{core}_\sigma(S)$, it follows from 4.4 that $p \in C(S \setminus E)$. In either case $C(S) \setminus E \subset C(S \setminus E)$. $\square$

Let $S \subset X$ with $|S| \geqslant \sigma$, By $Cl_\sigma(S)$ we denote the set of all cluster extreme sets in S with σ elements, i.e., $E \in Cl_\sigma(S)$ iff $C(E) \cap C(S \setminus E) = \phi$ and $|S| = \sigma$. Note that $Cl_\sigma(S) \subseteq E_\sigma(S)$.

THEOREM 4.6. Let $(X,C) = (R^d, conv)$ and $S \subseteq R^d$ with $|S| = m \geqslant \sigma$. Then the following holds

1. $core_\sigma(S) = \cap\{conv(S \setminus E) \mid E \in Cl_\sigma(S)\}$

2. $n*(\sigma,s) = |Cl_\sigma(S)|$ is the number of <u>non</u>-Radon 2-partitions of S with one component consisting of σ elements

Proof. (1) According to 4.4 we may restrict the proof to extreme sets. Let $M \subseteq S$ be an extreme set with $|M| = \sigma$. M can be partitioned into pairwise disjoint sets, say $M = M_1 \cup \cdots \cup M_k$, such that each M_i is contained in an extreme cluster of S. As $conv(S \setminus M) = conv(S \setminus (\cup_{i=1}^k M_i)) = conv(\cap_{i=1}^k (S \setminus M_i)) = \cap_{i=1}^k conv(S \setminus M_i)$ it follows that $\cap\{conv(S \setminus E) \mid E \in Cl_\sigma(S)\} \subseteq \cap\{conv(S \setminus M) \mid M \in E_\sigma(S)\} = core_\sigma(S)$. Hence, (1) holds. (2) is a direct consequence of (1) and the definition of extreme cluster set. $\square$

For the number $n*(\sigma,s) = |Cl_\sigma(S)|$ with $|S| = m \geqslant \sigma \geqslant 1$ in 4.6 it follows that:

$$n*(\sigma,s) \leqslant \binom{m}{\sigma} - r_\sigma$$

where r_σ is the σ-component of the *Radon vector*

$$(r_1, \ldots, r_{[m/2]})$$

of S; if $|S| = i$, then r_i is the number of Radon 2-partitions of S, say $\{S_1, S_2\}$, such that $|S_1| = \sigma$ and $|S_2| = i - \sigma$. For a survey of results on Radon vectors, see Eckhoff [9]. It is still an open problem to give an intrinsic characterization of the Radon vector $(r_1, \ldots, r_{[m/2]})$.

PROBLEM 11. Find an intrinsic characterization of the number $n*(\sigma,S)$, where $|S| = m \geqslant \sigma \geqslant 1$. Is $n*(\sigma,s) = \binom{m}{\sigma} - r_\sigma$ if S is in general position? Note that the first part of this problem is equivalent to [9] Problem 3.1.

THEOREM 4.7. For any aligned space (X,C) the following holds:

$$\max\{n \mid h(\sigma) = h(\sigma,n)\} \leqslant h(1)$$

Proof. Direct consequence of 4.1. □

THEOREM 4.8. For the ordinary aligned space with dimension d it follows that

$$\max\{n \mid h(\sigma) = h(\sigma,n)\} = h(1) = d + 1$$

Proof. Direct consequence of 4.3 and 4.7. □

For (R^d, conv) Reay [21] also has introduced the number $T(d,\tau,n,k)$; it is the (τ,n)-Radon number with the additional property that

$$\dim[\text{conv}(S_{i_1}) \cap \cdots \cap \text{conv}(S_{i_n})] \geqslant k$$

for each $i_1,\ldots,i_n \in \{1,\ldots,n\}$. Note that $T(d,\sigma,n,0)$ is the (τ,n)-Radon number of (R^d, conv).

Similarly, we define the number $H(d,\sigma,n,k)$ as the (σ,n)-Helly number of (R^d, conv) with the additional property that

$$\dim[\text{conv}(S \setminus M_i) \cap \cdots \cap \text{conv}(S \setminus M_n)] \geqslant k$$

for each $M_1,\ldots,M_n$ in S with $|M_i| \leqslant \sigma$ and each $i \in \{1,\ldots,n\}$. Note that $H(d,\sigma,n,0)$ is the (σ,n)-Helly number of (R^d, conv).

The following theorem is another generalization of Levi's theorem; the proof is similar to the previous generalizations; see Theorem 4.2.

THEOREM 4.9. $H(d,\sigma,n,k) \leqslant T(d,\sigma + 1,n,k) - 1$.

PROBLEM 12. Calculate $H(d,\sigma,n,k)$ under certain independence conditions, e.g., k-independence; see e.g., [6]. Does equality hold in 4.6?

In Larman [17] Radon partitions are studied for sets in R^d where some of its points are "stolen." Along this line the Larman-Radon and Larman-Helly numbers are introduced below.

Let ν be an integer $\geqslant 0$. The (τ,ν)-*Larman-Radon* number $LR(\tau,\nu)$ of an aligned space (X,C) is the infimum of all positive integers k for which it is true that for each set S in X with $|S| \geqslant k$ and each set T in S with $|T| \leqslant \nu$: there exists a τ-partition $S = S_i \cup \cdots \cup S_\tau$ such that

$$C(S_1 \setminus T) \cap \cdots \cap C(S \setminus T) \neq \phi$$

For $(R^d, conv)$ the number $LR(d,\tau,n,k,\nu)$ may be defined similar to the number $T(d,\tau,n,k)$.

The (σ,ν)-*Larman-Helly number* $LH(\sigma,\nu)$ of an aligned space (X,C) is defined as the infimum of all nonnegative integers k such that each S in X with $|S| \geqslant k + 1$ and each T in S with $|T| \leqslant \nu$ have the property that

$$\cap \{ C(S \setminus (M \setminus T)) \mid M \subseteq S, |M| \leqslant \sigma \} \neq \phi$$

The number $LH(d,\sigma,n,k,\nu)$ can be defined similar to the numbers $h(d,\sigma,n,k)$ and $LR(d,\tau,n,k,\nu)$.

PROBLEM 13.

(a) Is $LH(\sigma,\nu) \leqslant LR(\sigma + 1,\nu) - 1$ in an arbitrary aligned space?

(b) Is $LH(d,\sigma,n,k,\nu) = LR(d,\sigma + 1,n,k,\nu) - 1$ in $(R^d, conv)$?

(c) Calculate $LT(d,\sigma,n,k,\nu)$ under certain independence condition, e.g., k-independence.

REFERENCES

1. C. Berge and P. Duchet, Une généralisation du Théorème de Gilmore, *Cahiers Centre Etud. Rech. Opér.* 17(1975), 117-123.

2. P. M. Cohn, *Universal Algebra*, Harper and Row, New York, 1965.

3. L. Danzer, B. Grünbaum, and V. Klee, Helly's theorems and its relatives, *Proc. of Sympos.*, *Pure Math. VII*, Amer. Math. Soc., 1963, 101-180.

4. E. Degreef, The convex sum space and direct sum space, Report CSCOTW/123, Vrÿe Univ., Brussels, 1979.

5. J. P. Doignon, Convexity in crystallographical lattices, *J. of Geometry* 3(1973), 71-85.

6. J. P. Doignon, Radon partitions with k-dimensional intersections, to be published in *Journal of the London Math. Soc.*

7. J. P. Doignon, J. R. Reay, and G. Sierksma, A Tverberg-type generalization of the Helly number of a convexity space, to be published in *Archiv der Mathematik*.

8. J. Eckhoff, Der Satz von Radon in konvexen produktstrukturen I, *Monatshefte für Math.* 72(1968), 303–314.

9. J. Eckhoff, Radon's theorem revisited, *Proc. des Geometrie Symp.*, *Siegen 1978*, Birkhauser Verlag, Basel.

10. P. C. Hammer, Extended topology: Domain finiteness, *Indag. Math.* 25 (1963), 200–212.

11. B. Grünbaum, The dimension of intersections of convex sets, *Pacific J. Math.* 12(1962), 197–202.

12. R. E. Jamison, A general theory of convexity, Doct. Diss., Univ. of washington, Seattle, Wash., 1974.

13. R. E. Jamison-Waldner, Partition numbers for trees and ordered sets, Report 500, Tech. Hochschule, Darmstadt, 1979.

14. M. Katchalski, The dimension of intersections of convex sets, *Israël J. Math.* 10(1971), 465–470.

15. M. Katchalski, The dimension of intersections of convex sets II, *Israël J. Math.* 26(1977), 209–213.

16. D. C. Kay and E. W. Womble, Axiomatic convexity theory and the relationships between the Carathéodory, Helly, and Radon numbers, *Pacific J. of Math.* (2), 38(1971), 471–485.

17. D. G. Larman, On sets projectively equivalent to the vertices of a convex polytope, *Bull. London Math. Soc.* 4(1972), 6–12.

18. F. W. Levi, On Helly's theorem and the axioms of convexity, *J. Indian Math. Soc.* 15(1951), 65–76.

19. B. B. Peterson, The geometry of Radon's theorem, *Amer. Math. Monthly* 79(1972), 949–963.

20. J. R. Reay, An extension of Radon's theorem, *Illinois J. Math.* 12(1968), 184–189.

21. J. R. Reay, Several generalizations of Tverberg's theorem, *Israël J. Math.* (3), 34(1979), 238–244.

22. G. Sierksma, Carathéodory and Helly numbers of convex product structures, *Pacific J. Math.* (1), 61(1975), 275–282.

23. G. Sierksma, Axiomatic convexity theory and the convex product space, Doct. Diss., Univ. of Groningen, 1976.

24. G. Sierksma, Relationships between Carathéodory, Helly, Radon, and exchange numbers of convexity spaces, *Nieuw Archief voor Wisk.* (3), XXV (1977), 115–132.

25. G. Sierksma, Convexity on unions of sets, *Compositio Mathematica*, (3), 42(1981), 391–400.

26. H. Tverberg, A generalization of Radon's theorem, *J. London Math. Soc.* 41(1960), 123–128.

UNIMORPHIES OF SUBSETS OF HAUSDORFF LOCALLY CONVEX VECTOR SPACES

René Fourneau

Unité de Mathématiques
Institut Supérieur Industriel Liégeois
Liège, Belgium

1. INTRODUCTION, NOTATION, AND TERMINOLOGY

We prove that, in many cases, unimorphisms (i.e., generalized isometries) of subsets of Hausdorff locally convex vector spaces are affine.

An interesting theorem of R. Schneider [9] asserts that the onto isometries of the space $K(R^d)$ of all nonvoid compact convex subsets of R^d, metrized by the Hausdorff metric, are generated by the isometries of R^d: every isometry ϕ: $K(R^d) \to K(R^d)$ is such that

$$\phi(K) = \{f(x) : x \in K\} \qquad \forall K \in K(R^d)$$

where f is a rigid motion of R^d.

The author was surprised, looking at the proof of this theorem, to see the amount of properties of R^d which were used. In fact, $K(R^d)$ can be embedded isometrically and semilinearly in a normed vector space; and we think that a generalization of the Mazur-Ulam theorem with the help of 6.2 of [3] must be the correct tool for this proof, giving in the same time a way of generalization of the theorem of Schneider. Other theorems of P. Gruber [4, 5] could receive the same treatment.

The difficulty is of course to prove a Mazur-Ulam-like theorem for (at least) convex cones. We shall present here the results obtained in trying to find this theorem.

All vector spaces we shall deal with are real vector spaces. We shall describe the topology of a Hausdorff locally convex topological vector space by a (upward directed and separating) system of seminorms.

The closed semiball with center a and radius $r > 0$ is the set

$$\{x \in V : p(x - a) \leqslant r\}$$

where p belongs to the system of seminorms of V.

We shall denote by V* the topological dual of the topological vector space V and by $\sigma(V, V^*)$ the weak topology of V associated with this duality.

By a *strictly convex normed vector space*, we mean a normed vector space the unit ball of which is strictly convex, that is, such that the unit sphere does not include nonsingleton segments.

If V_1 and V_2 are real vector spaces, if $S_1 \subset V_1$ and $S_2 \subset V_2$, a map $\phi: S_1 \to S_2$ is said to be *semiaffine* if, for every $x_1, x_2 \in S_1$,

$$\phi[\lambda x_1 + (1 - \lambda)x_2] = \lambda\phi(x_1) + (1 - \lambda)\phi(x_2)$$

for every $\lambda \in R$ such that $\lambda x_1 + (1 - \lambda)x_2 \in S_1$.

A *generating convex cone* of a vector space V is a convex cone C such that $C - C = V$.

In addition, we shall use the notation and terminology of [1].

2. UNIMORPHIES OF LOCALLY CONVEX VECTOR SPACES

Let V_1 and V_2 be Hausdorff locally convex vector spaces, the topologies of which being generated by the systems of seminorms P_1 and P_2 respectively.

Let S_1 [resp. S_2] be a subset of V_1 [resp. V_2]. A *unimorphism* from S_1 to S_2 is a couple (ϕ, F), the first component of which is an injection $\phi: S_1 \to S_2$ and the second is a bijection $F : P_1 \to P_2$ such that for any $p \in P$

$$F(p)[\phi(x) - \phi(y)] = p(x - y) \qquad \forall x, y \in S$$

Obviously, if ϕ is a unimorphism from S to S', ϕ is a uniform homeomorphism from S into S' endowed with the uniformities induced on them by that of V_1 and V_2, respectively.

3. A MANKIEWICZ-TYPE THEOREM FOR UNIMORPHIES

The special case of the following theorem for normed spaces has been proved by P. Mankiewicz ([6], Th. 2, p. 368). This author has given a lot of very nice generalizations in other directions but apparently missed our theorem for technical reasons ([7, 8]).

It is possible to deduce our version of Mankiewicz's theorem using a quotient technique, but the direct proof is nece and useful for generalizations.

THEOREM 3.1 If (ϕ, F) is a unimorphism from an open connected subset S_1 of a Hausdorff locally convex vector space V_1 onto an open subset S_2 of a Hausdorff locally convex vector space V_2, there is a unique extension (Φ, F) of (ϕ, F) which is a translate of a linear unimorphism from V_1 onto V_2.

Proof. Assume that $0 \in \overset{\circ}{S}$ and that $\phi(0) = 0$. Let B be a closed p-semiball of radius $r > 0$ included in S and let d be the semidistance associated with p. Assume for simplicity of notation that the center of B is 0.

The radius r will be chosen small enough in such a way that

$$\{y \in V_2 : F(p)(y) \leqslant r\} \subset S_2$$

When r has been chosen this way, we shall say that B is *sufficiently small.*

We shall first show that, for any $x \in B$ and $\lambda \geqslant 0$ such that $\lambda x \in B$,

$$\phi(\lambda x) = \lambda \phi(x)$$

In fact, it suffices to prove this equality for any $x \in B$ and $\lambda \in [0,1]$. The proof is a modification of that of Lemma 1, p. 142 of [2].

Let $p' \in P_1$ and let d' be the semidistance associated with p'. Set

$$E_1 = \{y \in B : d'(0,y) = d'(y,x) = \tfrac{1}{2} d'(0,x) \text{ and } d(x,y) \leqslant r\}$$

and, if E_n has been defined, let $D'(E_n)$ be the d'-diameter of E_n and set

$$E_{n+1} = \{y \in E_n : d'(y,x) \leqslant \frac{D'(E_n)}{2}, \qquad \forall z \in E_n\}$$

It is easy to see that if $E_{n+1} \neq \phi$, $D'(E_{n+1}) \leqslant \frac{D'(E_n)}{2}$.

Let T be the map defined by

$$Ty = x - y, \qquad \forall y \in V_1$$

Let $y \in E_1$,

$$d'(Ty,0) = p'(x - y) = d'(x,y)$$

and

$$d'(Ty,x) = p'(x - y - x) = d'(0,y)$$

hence $Ty \in E_1$, since

$$p(Ty) = p(x - y) = d(x,y) \leqslant r$$

Suppose now that z and $Tz \in E_n$ and $y \in E_{n+1}$; then

$$d'(Ty,z) = p'(x - y - z) = p'(x - z - y) = d'(Tz,y) \leqslant \frac{D'(E_n)}{2}$$

so $Ty \in E_{n+1}$. Therefore, by induction, $T(E_n) \subset E_n$ for every $n \in N$.
 Now, let $y \in E_n$,

$$d'(y,\frac{x}{2}) = p'(y - \frac{x}{2}) = \frac{1}{2} p'(x - y - y) = \frac{d'(Ty,y)}{2} \leqslant \frac{D'(E_n)}{2}$$

hence $\frac{x}{2} \in E_{n+1}$ if $\frac{x}{2} \in E_n$.
 Now since

$$d(x,\frac{x}{2}) = p(x - \frac{x}{2}) = p(x - \frac{x}{2}) = p(\frac{x}{2}) = \frac{1}{2} p(x) < r$$

and

$$d'(0,\frac{x}{2}) = d'(\frac{x}{2},x) = \frac{1}{2} d'(0,x)$$

$\frac{x}{2} \in E_1$, and therefore, by induction,

$$\frac{x}{2} \in \bigcap_{n=1}^{\infty} E_n = M_p,$$

the d'-diameter of M_p, is obviously zero.

Since $\frac{x}{2} \in M_{p'}$ for every $p' \in P_1$,

$$\frac{x}{2} \in \underset{p \in P_1}{\cap} M_{p'} = M$$

where M has d'-diameter zero for every $p' \in P_1$, hence is reduced to a point:
$M = \{\frac{x}{2}\}$.

The image of B under ϕ is the semiball

$$\{x \in V_2 \ : \ F(p)(x) \leqslant r\}$$

and it is readily seen that, for every $p_1' \in P_1$

$$M_{F(p')} = \phi(M_{p'})$$

where $M_{F(p')}$ has been constructed from $\phi(x)$ like above. Therefore,

$$\phi\left(\frac{x}{2}\right) = \frac{\phi(x)}{2}$$

In fact, we have shown that, if x_1 is the center of a sufficiently small
closed p-semiball included in S and if x_2 is a point of this semiball, the
image under ϕ of the midpoint of $[x_1 : x_2]$ is the midpoint of $[\phi(x_1) : \phi(x_2)]$.

Now, starting with 0 and $\frac{x}{2}$ for new x_1 and x_2 respectively, one gets
$\phi(\frac{x}{4}) = \frac{1}{4} \phi(x)$ and, since the semiball

$$\{y \ : \ p(\frac{x}{2} - y) \leqslant \frac{r}{2}\}$$

is included in B, the use of $\frac{x}{2}$ for x_1 and x for s_2 gives

$$\phi(\frac{3x}{4}) = \frac{3}{4} \phi(x)$$

Using this process inductively gives

$$\phi(\lambda x) = \lambda \phi(x)$$

for every dyadic rational $\lambda \in [0,1]$, which proves the equality for ϕ is continuous.

We can now define $\Phi: V_1 \rightarrow V_2$ by

$$\Phi(x) = \frac{p(x) + \varepsilon}{r} \left[\frac{r}{p(x) + \varepsilon} \cdot x \right]$$

where ε is a fixed positive constant.

The mapping Φ is an extension of $\phi_{|B}$ since, for any $x \in B$,

$$p\left(\frac{r}{p(x) + \varepsilon} \cdot r \right) < r \quad \text{and} \quad \frac{p(x) + \varepsilon}{r} \frac{r}{p(x) + \varepsilon} \cdot x = x$$

hence both x and $\dfrac{r}{p(x) + \varepsilon} \cdot x$ belong to B and

$$\Phi(x) = \frac{p(x) + \varepsilon}{r} \phi\left(\frac{r}{p(x) + \varepsilon} \cdot x \right) = \phi(x)$$

Let us show that Φ is a unimorphism. For any x, $y \in V_1$ and $p' \in P_1$, if we assume $p(y) \leqslant p(x)$,

$$F(p')[\Phi(x) - \Phi(y)] = F(p') \left[\frac{p(x) + \varepsilon}{r} \phi\left(\frac{r}{p(x) + \varepsilon} \cdot x \right) - \frac{p(y) + \varepsilon}{r} \phi\left(\frac{r}{p(y) + \varepsilon} \cdot y \right) \right]$$

$$= \frac{p(x) + \varepsilon}{r} F(p') \left[\phi\left(\frac{r}{p(x) + \varepsilon} \cdot x \right) - \frac{p(y) + \varepsilon}{p(x) + \varepsilon} \phi\left(\frac{r}{p(y) + \varepsilon} \cdot y \right) \right]$$

$$\overset{(*)}{=} \frac{p(x) + \varepsilon}{r} F(p') \left[\phi\left(\frac{r}{p(x) + \varepsilon} \cdot x \right) - \phi\left(\frac{r}{p(x) + \varepsilon} \cdot y \right) \right]$$

$$= \frac{p(x) + \varepsilon}{r} p' \left[\frac{r}{p(x) + \varepsilon} \cdot x - \frac{r}{p(x) + \varepsilon} \cdot y \right]$$

$$= p'(x - y)$$

where we used, in $(*)$, the fact that $\dfrac{p(y) + \varepsilon}{p(x) + \varepsilon} \leqslant 1$.

The Mazur-Ulam theorem for unimorphies ([2], Corollary 1, p. 143) asserts then that Φ is linear. We can therefore conclude that $\phi_{|B}$ is the restriction of a linear map. Moreover, any linear extension of $\phi_{|B}$ coincides with Φ since

$$\phi^*(x) = \phi^* \left[\frac{p(x) + \varepsilon}{r} \frac{r}{p(x) + \varepsilon} \cdot x \right] = \frac{p(x) + \varepsilon}{r} \phi^* \left[\frac{r}{p(x) + \varepsilon} \cdot x \right]$$

$$= \frac{p(x) + \varepsilon}{r} \phi\left(\frac{r}{p(x) + \varepsilon} \cdot x \right) = \Phi(x)$$

for any $x \in V_1$. In particular, Φ is independent of $\varepsilon > 0$.

If B_1, B_2 are sufficiently small semiballs included in S such that $\overset{\circ}{B}_1 \cap \overset{\circ}{B}_2 \neq \phi$, the extensions constructed as above from B_1 and B_2 must coincide since $B_1 \cap B_2 \supset B_3$ and $\phi|_{B_3}$ has a unique extension. The connectivity of S can then be used to show that Φ is an extension of ϕ and that Φ is the unique extension of ϕ which is linear. $\square$

4. WEAK UNIMORPHIES

Using the preceding theorem and a quotient technique, one can prove the following theorem on weak onto unimorphies. We give a direct proof, for it can be used to get generalizations (see Section 5).

THEOREM 4.1. Let V_1 and V_2 be locally convex topological vector spaces, V_1 being endowed with its weak topology. Every unimorphism ϕ from a convex subset C_1 of V_1 onto a convex subset C_2 of V_2 is semiaffine.

Proof. Let x_1, $x_2 \in C_1$. For every linear form f on E_1, if

$$m = \frac{x_1 + x_2}{2} \quad \text{and} \quad \alpha_f = \frac{1}{2} |f(x_1) - f(x_2)|$$

$$m \in \{x \in C_1 : |f(x) - f(x_1)| = \alpha_f\} \cap \{x : |f(x) - f(x_2)| = \alpha_f\}$$

$$= \left\{x : f(x) = \frac{f(x_1) + f(x_2)}{2}\right\}$$

Therefore,

$$m \in \underset{f \in E_1^*}{\cap} [\{x \in C_1 : |f(x) - f(x_1)| = \alpha_f\} \cap \{x : |f(x) - f(x_2)| = \alpha_f\}]$$

$$= \underset{f \in E_1^*}{\cap} \left\{x \in C_1 : f(x) = \frac{f(x_1) + f(x_2)}{2}\right\}$$

Since the last set is a singleton, we must have

$$m = \underset{f \in E_1^*}{\cap} [\{x \in C_1 : |f(x) - f(x_1)| = \alpha_f\} \cap \{x : |f(x) - f(x_2)| = \alpha_f\}]$$

Moreover, it is easily verified that

$$m \in \{x : \sup_{i=1,\ldots,n} |f_i(x) - f_i(x_1)| = \frac{1}{2} \sup_{i=1,\ldots,n} |f_i(x_1) - f_i(x_2)|\}$$

$$\cap \{x : \sup_{i=1,\ldots,n} |f_i(x) - f_i(x_2)| = \frac{1}{2} \sup_{i=1,\ldots,n} |f_i(x_1) - f_i(x_2)|\}$$

where $f_1,\ldots,f_n \in E_1^*$, hence

$$m = \bigcap_{p \in S_1} [\{x \in C_1 : p(x - x_1) = \frac{1}{2} p(x_1 - x_2)\}]$$

where S_1 is the system of seminorms of $\sigma(E_1, E_1^*)$.

From there on,

$$\phi(m) = \bigcap_{p \in S_1} [\phi\{x : p(x - x_1) = \frac{1}{2} p(x_1 - x_2)\}$$

$$\cap \phi\{x : p(x - x_2) = \frac{1}{2} p(x_1 - x_2)\}]$$

$$= \bigcap_{p \in S_1} [\{y \in C_2 : F(p)(y - \phi(x_1)) = \frac{1}{2} F(p)(\phi(x_1) - \phi(x_2))\}$$

$$\cap \{y \in C_2 : F(p)(y - \phi(x_2)) = \frac{1}{2} F(p)(\phi(x_1) - \phi(x_2))\}]$$

$$= \bigcap_{p \in P_2} [\{y \in C_2 : p(y - \phi(x_1)) = \frac{1}{2} p(\phi(x_1) - \phi(x_2))\}$$

$$\cap \{y \in C_2 : p(y - \phi(x_2)) = \frac{1}{2} p(\phi(x_1) - \phi(x_2))\}]$$

But, for every $p \in P_2$,

$$p\left(\frac{\phi(x_1) + \phi(x_2)}{2} - \phi(x_1)\right) = p\left(\frac{\phi(x_2) - \phi(x_1)}{2}\right) = \frac{1}{2} p(\phi(x_2) - \phi(x_1))$$

and, likewise,

$$p\left(\frac{\phi(x_1) + \phi(x_2)}{2} - \phi(x_2)\right) = \frac{1}{2} p(\phi(x_1) - \phi(x_2))$$

The equality

$$\phi\left(\frac{x_1 + x_2}{2}\right) = \frac{\phi(x_1) + \phi(x_2)}{2}$$

is thence plain, since $\dfrac{\phi(x_1) + \phi(x_2)}{2} \in C_2$.

Using the continuity of ϕ and dyadic approximation, one can show that ϕ is semiaffine:

$$\phi(\lambda x_1 + (1 - \lambda)x_2) = \lambda\phi(x_1) + (1 - \lambda)\phi(x_2)$$

for every x_1, $x_2 \in C_1$ and every $\lambda \in R$ such that $\lambda x_1 + (1 - \lambda)x_2 \in C_2$. $\square$

THEOREM 4.2. Let V_1 and V_2 be locally convex topological vector spaces endowed with their weak topology. Every unimorphism ϕ from a convex subset C_1 of V_1 into a subset C_2 of V_2 is semiaffine and $\phi(C_1)$ is convex.

The proof is a slight modification of that of 4.1.

5. UNIMORPHIES OF SPECIAL SPACES

LEMMA 5.1. If V_1 is a Hausdorff locally convex vector space, whose system of seminorms is P, such that for every two points x_1, $x_2 \in V_1$ the set

$$\bigcap_{p \in P} \{x \in V_1 : p(x - x_1) = \frac{1}{2} p(x_1 - x_2)\} \cap \{x \in V_1 : p(x - x_2) =$$

$$\frac{1}{2} p(x_1 - x_2)\}$$

is a singleton, every unimorphism from a convex subset of V_1 onto a convex subset of another Hausdorff locally convex vector space V_2 is semiaffine.

If both the systems of seminorms of V_1 and V_2 enjoy this property, every unimorphism from a convex subset C_1 of V_1 into a subset of V_2 is semiaffine and $\phi(C_1)$ is convex.

The proofs are trivial adaptations of 4.1 and 4.2, respectively.

The above lemma is rather trivial but is useful for proving certain affineness theorems for unimorphisms. For example, one can deduce the following statement:

If ϕ is a unimorphism from a convex subset of a strictly convex normed space onto a convex subset of a normed space, ϕ is semiaffine.

If ϕ is a unimorphism from a convex subset C_1 of a strictly convex normed space into a subset of a strictly convex normed space is semiaffine and $\phi(C_1)$ is convex.

6. A CONJECTURE

The preceding theorems lead to the following conjecture:

If ϕ is a unimorphism from a generating convex cone C_1 of a Hausdorff locally convex vector space V_1 onto a generating convex cone C_2 of a Hausdorff locally convex vector space V_2, ϕ is semiaffine.

REFERENCES

1. J. Bair and R. Fourneau, *Etude géométrique des espaces vectoriels--Une introduction, Lecture Notes in Math.*, 489, Springer-Verlag, Berlin, 1975.

2. M. M. Day, *Normed linear spaces*, 3rd, ed., *Ergeb. der Math.*, Band 21, Springer-Verlag, Berlin, 1973.

3. R. Fourneau, Isomorphismes de lattis et de demi-groupes ǎ opérateurs de fermés convexes, *Bull. Soc. Roy. Sc. Liége* 45(1976), 169-174.

4. P. Gruber, Isometries of the space of convex bodies of E, *Mathematika* 25(1978), 270-278.

5. P. Gruber and G. Lettl, Isometries of the space of convex bodies in euclidean space, *Bull. London Math. Soc.* 12(1980), 455-462.

6. P. Mankiewicz, On extensions of isometries in normed linear spaces, *Bull. Acad. Pol. Sc., Ser. Sc. Math., Astr. et Phys.* 20(1972), 367-371.

7. P. Mankiewicz, On isometries in linear metric spaces, *Studia Math.* 55 (1976), 163-173.

8. P. Mankiewicz, Fat equicontinuous groups of homeomorphisms of linear topological spaces and their applications to the problem of isometries in linear metric spaces, *Studia Math.* 64(1979), 13-23.

9. R. Schneider, Isometries des Raumes der Konvexen Körper, *Coll. Math.* 33 (1975), 219-224.

TILING R^d BY TRANSLATES OF THE ORTHANTS

Jim Lawrence

Department of Mathematics
University of Kentucky
Lexington, Kentucky

1. INTRODUCTION

A collection K of congruent cubes is said to *tile* R^d if

(i) $\cup K = R^d$, and

(ii) Each pair of cubes in K have disjoint interiors,

and we will call K a *tiling* of R^d. Such a tiling is said to be a *lattice tiling* if there is a nonsingular $d \times d$ matrix A such that $K = \{I^d + Az : z \in Z^d\}$ (where $I^d = \{x = (x_1, \ldots, x_d) \in R^d : 0 \leqslant x_i \leqslant 1, \text{ for } 1 \leqslant i \leqslant d\}$ and $Z^d = \{x \in R^d : x_i \text{ is an integer, for } 1 \leqslant i \leqslant d\}$).

In [3], Minkowski conjectured that, in any lattice tiling K of R^d by congruent cubes, some pair of the cubes in K share a common $(d - 1)$-dimensional face (i.e., a facet). (Such a pair of cubes are "back-to-back.") This was proven by Hajós in [1], about forty years later, but in the meantime Keller, in [2], had extended Minkowski's conjecture to arbitrary tilings by congruent cubes, and this conjecture is still unsettled (for $d > 6$). In 1942, Perron [4] proved Keller's conjecture for $d \leqslant 6$. He also formulated a somewhat simpler-looking conjecture (involving only 2^d cubes in R^d, instead of an infinite family), which he proved is equivalent to Keller's.

For more recent papers on these problems, see Stein [6] and Robinson [5].

In this paper we formulate yet another conjecture, which is equivalent to Keller's conjecture. This one is graph-theoretic; namely:

If the edge-set E of the complete graph with 2^d vertices is written as the union of those of d bipartite graphs, $E = E_1 \cup \cdots \cup E_d$, then there is an edge $e \in E_i$, for some i, such that $E_j \cup \{e\}$ contains the edge-set of an odd cycle, for each $j \neq i$.

The proof of the equivalence of this to Keller's conjecture forms the rest of the paper.

2. PROOF OF EQUIVALENCE

By a *box* in R^d we mean a set B of the form $B = B_1 \times \cdots \times B_d$, where each B_i is a connected, nonempty subset of R. (The set B_i can be a single point, or it can be an interval, finite or infinite, not necessarily open, or closed.) The intersection of two boxes is either empty or another box. Indeed, if $B^1 = B_1^1 \times \cdots \times B_d^1$ and $B^2 = B_1^2 \times \cdots \times B_d^2$, then $B^1 \cap B^2 = (B_1^1 \cap B_1^2) \times \cdots \times (B_d^1 \cap B_d^2)$.

Consider the problem of covering $I^d = I \times \cdots \times I$ (where $I = [0,1]$) by pairwise disjoint boxes contained in I^d. Call a box $B = B_1 \times \cdots \times B_d \subseteq I^d$ *proper* if for no i is $B_i = I$. How many proper boxes are required to cover I^d? At least 2^d are required, since no two of the 2^d vertices of I^d are in any proper box. There are several ways to cover I^d by 2^d pairwise disjoint boxes. We will say that a collection of boxes *packs* I^d if their union is I^d and they are pairwise disjoint.

The following is a variant of Perron's conjecture. We will show that the graph-theoretic conjecture stated in Section 1 is equivalent to it.

CONJECTURE. If the proper boxes $B^k = B_1^k \times \cdots \times B_d^k$, for $1 \leq k \leq 2^d$, pack I^d, then there are boxes B^{k_1} and B^{k_2} and j with $1 \leq j \leq d$ for which

$$B_i^{k_1} = B_i^{k_2} \text{ if } i \neq j, \text{ and}$$

$$B_j^{k_1} = I \sim B_j^{k_2}$$

(The boxes B^{k_1} and B^{k_2} are, in a sense, back-to-back.)

In Theorem 1 we give a simple combinational criterion for 2^d proper boxes to pack I^d. First we need a lemma.

LEMMA. For $1 \leqslant k \leqslant 2^d$, let $B^k = B_1^k \times \cdots \times B_d^k$ be a proper box. If these boxes pack I^d, then so do the boxes $\bar{B}^k = (I \sim B_1^k) \times B_2^k \times \cdots \times B_d^k$.

Proof. Since each vertex of I^d is contained in some box, and no box contains two vertices, each of the 2^d boxes must contain some vertex. Therefore each set B_i^k contains 0 or 1. Suppose $a = (a_1, \ldots, a_d)$ is in I^d. Let $L = \{(x, a_2, \ldots, a_d) : 0 \leqslant x \leqslant 1\}$. The set L must intersect exactly two boxes, the one, say B^m, that contains $(0, a_2, \ldots, a_d)$, and the one, B^n, that contains $(1, a_2, \ldots, a_d)$. The vector a is in one of these. If it is in B^m, then it is also in $\bar{B}^n$. If it is in B^n, then it is also in $\bar{B}^m$. It follows that the union of the boxes $\bar{B}^k$ is I^d.

We must show that the boxes $\bar{B}^k$ are pairwise disjoint. Suppose $1 \leqslant m < n \leqslant 2^d$. The boxes B^m and B^n are disjoint, so there is i with $B_i^m \cap B_i^n = \phi$. If $i \neq 1$, then $\bar{B}^m$ and $\bar{B}^n$ are disjoint. Suppose the only i with $B_i^m \cap B_i^n = \phi$ is $i = 1$. Choose $b_j \in B_j^m \cap B_j^n$, for $j = 2, \ldots, d$. Let $L' = \{(x, b_2, \ldots, b_d) : 0 \leqslant x \leqslant 1\}$. Then L' is the disjoint union of $B^m \cap L'$ and $B^n \cap L'$. It follows that B_1^m and B_1^n are complementary, and $\bar{B}^m \cap \bar{B}^n = \phi$. $\square$

Given a box $B = B_1 \times \cdots \times B_d$ and a set $S \subseteq \{1, 2, \ldots, d\}$, let $B(S)$ be the box $B_1(S) \times \cdots \times B_d(S)$, where

$$B_i(S) = \begin{cases} B_i & \text{if } i \notin S \\ I \sim B_i & \text{if } i \in S \end{cases}$$

The preceding lemma states that if boxes B^k pack I^d, so do the boxes $B^k(\{1\})$. Clearly we may replace "1" by any other index, or, indeed, replace $\{1\}$ by any set $S \subseteq \{1, \ldots, d\}$, here, so that the boxes $B^k(S)$ also pack I^d, for any such set S.

THEOREM 1. For $1 \leqslant k \leqslant 2^d$, let $B^k = B_1^k \times \cdots \times B_d^k$ be a proper box. These 2^d boxes pack I^d if and only if, for each pair m and n with $1 \leqslant m < n \leqslant 2^d$, there is an index i for which $B_i^m = I \sim B_i^n$.

Proof. Suppose the boxes pack I^d. Suppose $1 \leqslant m < n \leqslant 2^d$. Let $S = \{j : B_j^m \cap B_n^n = \phi\}$. Since the boxes $B^k(S)$ also pack I^d, $B^m(S) \cap B^n(S) = \phi$. Then for some i, $B_i^m(S) \cap B_i^n(S) = \phi$. Clearly, $i \in S$, and, since both $B_i^m \cap B_i^n = \phi$ and $(I \sim B_i^m) \cap (I \sim B_i^n) = \phi$, $B_i^m = I \sim B_i^n$.

Suppose the 2^d boxes have the property that for each pair B^m and B^n of distinct boxes, there is an index i for which $B_i^m = I \sim B_i^n$. Clearly the boxes are pairwise disjoint. Suppose $a \in I^d$. We must show that a is in one of the boxes. For each box B^k, let $v(B^k) = (\varepsilon_1, \ldots, \varepsilon_d)$ be the vector with

$$\varepsilon_i = \varepsilon_i(B^k) = \begin{cases} 0 & \text{if } a_i \notin B_i^k \\ 1 & \text{if } a_i \in B_i^k \end{cases}$$

Then if $m \neq n$, $v(B_m) \neq v(B_n)$, for if i is such that $B_i^m = I \sim B_i^n$, $\varepsilon_i(B_m) \neq \varepsilon_i(B^n)$. It follows that v is injective, and, since the cardinality of the set of boxes is the same as that of the sequences of 0's and 1's, bijective. There is k with $v(B^k) = (1,1,\ldots,1)$, and $a \in B^k$. $\square$

THEOREM 2. The graph-theoretic conjecture (*) is equivalent to Perron's and Keller's conjectures.

Proof. Suppose the proper boxes B^k ($1 \leq k \leq 2^d$) pack I^d. Let $G = (V,E)$ be the complete graph with vertex set $V = \{1,\ldots,2^d\}$. For $1 \leq i \leq d$, let G_i be the graph (V,E_i), where the edge $e = \{m,n\}$ is in E_i if $B_i^m = I \sim B_i^n$. Clearly G_i is bipartite; indeed, each connected component of G_i is a complete bipartite graph. Since the boxes B^k pack I^d, $E = E_1 \cup \cdots \cup E_d$, by Theorem 1. If (*) is true, then there is an edge $e = \{m,n\}$ in some E_i, such that $E_j \cup \{e\}$ contains the edges of an odd cycle, for each $j \neq i$. Then $B_i^m = I \sim B_i^n$, and $B_j^m = B_j^n$ for $j \neq i$; i.e., (*) implies Perron's conjecture.

Suppose $G = (V,E)$ is the complete graph with vertex set $V = \{1,\ldots,2^d\}$, and $E = E_1 \cup \cdots \cup E_d$, where $G_i = (V,E_i)$ is a bipartite graph (for each i). We will describe how to construct a family of boxes B^k which pack I^d. For $1 \leq i \leq d$, assign to each connected component of the graph G_i a number α, $0 < \alpha < 1$, to be called the *width* of the component, subject only to the restriction that no two components of G_i have the same width. Since G_i is bipartite we may 2-color its vertices using, say, red and blue. Let this be done. Finally, let $B^k = B_i^k \times \cdots \times B_d^k$, where, if k is in the component of G_i which has width α, then

$$B_i^k = \begin{cases} [0,\alpha) & \text{if k is colored red in } G_i \\ [\alpha,1] & \text{if k is colored blue in } G_i \end{cases}$$

Suppose $1 \leqslant m < n \leqslant 2^d$. Since $E = E_1 \cup \cdots \cup E_d$, the edge $e = \{m,n\}$ is in E_i for some i, and, for this i, $B_i^m = I \sim B_i^n$. It follows, by Theorem 1, that the boxes B^k pack I^d.

If Perron's conjecture is true, then there are boxes B^m and B^n, and an index i, with $B_i^m = I \sim B_i^n$ and $B_j^m = B_j^n$ for $j \neq i$. Since $B_j^m = B_j^n$ for $j \neq i$, m and n are in the same component of G_j and of the same color in G_j, so that the edge $e = \{m,n\}$ is on an odd cycle with all its other edges in E_j. Also, since then clearly $e \notin E_j$ for $j \neq i$, $e \in E_i$. It follows that Perron's conjecture implies (*). $\square$

ACKNOWLEDGMENT

We were privileged to discuss this problem with Branko Grünbaum. Among other things, he pointed out reference [5] to us.

REFERENCES

1. G. Hajós, Über einfache und mehrfache Redeckung des n-dimensionalen Raumes mit einem Würfelgitter, *Math. Z.* 47(1942), 427-467.

2. O. Keller, Über die lückenlose Erfüllung des Raumes mit Würfeln, *J. Reine Angew. Math.* 163(1930), 231-248.

3. H. Minkowski, *Geometrie der Zahlen*, erste Lieferung, Leipzig, B. G. Teubner, 1896.

4. O. Perron, Über lückenlose Ausfüllung des n-dimensionalen Raumes durch kongruente Würfel, *Math. Z.* 46(1940), 1-26, 161-180.

5. R. M. Robinson, Multiple tilings of n-dimensional space by unit cubes, *Math. Z.* 166(1979), 225-264.

6. S. K. Stein, Algebraic tiling, *Amer. Math. Monthly*, 81(1974), 445-462.

EBERHARD'S THEOREM FOR 4-VALENT CONVEX 3-POLYTOPES

Joseph Malkevitch

Department of Mathematics
York College
Jamaica, New York

If p_k (or $p_k(P)$) denotes the number of faces of a convex 3-valent 3-polytope P (see [4] for standard terminology) then:

$$3p_3 + 2p_4 + p_5 = 12 + \sum_{k \geq 7} (k - 6)p_k \qquad (1)$$

This result is an easy consequence of Euler's formula. It is interesting to consider if 3-valent 3-polytopes can be constructed corresponding to each solution of (1) in nonnegative integers. This turns out not to be the case, as can easily be seen by considering the solution to (1), $p_3 = 5$, $p_9 = 1$. Noticing that (1) places no restriction on p_6, one is tempted to see if hexagons can be used together with other polygons whose numbers satisfy (1) to construct a 3-valent 3-polytope. More precisely, if p_3^*, p_4^*, p_5^*, p_7^*, ..., p_n^* is a nonnegative integer solution of (1), one would like to know if there exists a convex 3-valent 3-polytope P and a choice of p_6^* such that $p_k(P) = p_k^*$, for all values of k, $3 \leq k \leq n$, and $p_k(P) = 0$ for $k > n$. The answer to this question is given by Eberhard's Theorem [1, 4]:

THEOREM 1 (V. Eberhard). If p_3^*, p_4^*, p_5^*, p_7^*, ..., p_n^* is a nonnegative integer solution of (1), then there exists a nonnegative integer p_6^* and a convex 3-valent 3-polytope P such that $p_k(P) = p_k^*$ for k, $3 \leqslant k \leqslant n$, and $p_k(P) = 0$ for $k > n$.

For proofs see [1, 4, 7]. All proofs of this statement require careful "bookkeeping" in the construction of an appropriate plane, 3-connected 3-valent graph, on which Steinitz's Theorem [4] can be invoked, thereby guaranteeing the existence of the appropriate 3-polytope. Steinitz's Theorem states that necessary and sufficient conditions that a graph G be isomorphic to the graph of vertices and edges of a convex 3-polytope is that G be planar and 3-connected. The spirit of proof of an Eberhard type theorem is given by the much more transparent proof of a generalization of Eberhard's Theorem to obtain a 4-valent analogue, due to Branko Grünbaum.

THEOREM 2 (B. Grünbaum). If p_3^*, p_5^*, p_6^*, ..., p_n^* is a nonnegative integer solution of equation (2):

$$p_3 = 8 + \sum_{k \geqslant 5} (k - 4)p_k \tag{2}$$

then there exists a nonnegative integer p_4^* and a convex 4-valent 3-polytope P such that $p_k(P) = p_k^*$ for $3 \leqslant k \leqslant n$, and $p_k(P) = 0$ for $k > n$.

The proof of this result is very elegant and appears in [4]. Ernst Jucovic [7] gives another proof of this result and proves the following extension.

THEOREM 3 (E. Jucovic). For every sequence $(p_3^*, p_4^*, ..., p_n^*)$ satisfying (1) and with $p_4^* \geqslant \sum_{k=5}^{n} (3k - 10)p_k^*$, there exists a 3-polytope P with $p_k(P) = p_k^*$, $3 \leqslant k \leqslant n$.

T. C. Enns [2] has recently strengthened this theorem. The purpose of this note is to give a new proof of Theorem 2 and a result in the spirit of Theorem 3.

The proof will exploit an extension of an idea developed in [9, 11] and used to solve a problem in [5].

Let G be a plane k-valent (k = 3, 4, or 5) 2-connected graph in a fixed
embedding in the plane. By a *k-shell diagram of length m* is meant a plane
simple circuit of length m with (k - 2) short spikes into the interior and/or
exterior of points on the m circuit. Figure 1 shows two 3-shell diagrams of
length 6, for k = 3.

The exterior code of a 3-shell diagram is the number of outspikes at each
point of the diagram. The codes for the diagrams in Fig. 1 are (1,1,0,0,1,0)
and (1,1,0,1,1,0), respectively. Note we treat the code cyclically, and it
could be considered to start in any position. Two codes are the same if they
are identical when read in either cyclic order. The interior code of a shell
diagram is the number of inspikes at each point of the diagram. The interior
codes for the diagram in Fig. 1 are (0,0,1,1,0,1) and (0,0,1,0,0,1), respectively. If the interior code and exterior codes are identical, the 3-shell
diagram is called self-invertible or invertible. The diagram on the left in
Fig. 1 is invertible but the one on the right is not. The 3-shell diagram
on the left in Fig. 2 also has the property that it is replicable by a single
hexagon. This is shown in Fig. 2. After a single hexagon is adjoined, the
resulting shell diagram is the same as the original one.

The wiggled circuit is the new shell diagram, which has the same codes
as the original. To get a new proof of Theorem 2, we will use the fact that
Fig. 3 shows a 4-shell diagram of even length (which is invertible) and which
is replicable by a single 4-gon.

Using the existence of this 4-shell, we see that if there exists a planar
4-valent, 3-connected graph G which contains such a shell, then there also
exist 4-valent 3-connected graphs H_m with m 4-gons for all values of m which
exceed $p_4(G)$.

FIGURE 1

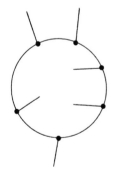 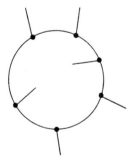

FIGURE 2

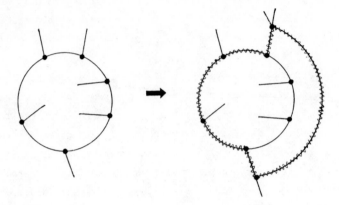

To show how to get a new proof of Theorem 2, we first show that the interior of a 4-shell of even length can have its interior completed with faces
which are only 3-gons and 4-gons. The construction is shown in Fig. 3. The
diagram on the right shows, with the dotted lines, how to increase the length
of the shell by 2, and create only new 4-gons.

For a solution of (2) with faces of length $\geqslant$ 5, we proceed as follows.
The general case, is done in the same fashion as the example. To create an
8-gon, 5-gon, and 6-gon see Fig. 4.

Note that for each k-gon, $(k \geqslant 5)$ $k - 4$ triangles are created, as required by (2). Furthermore, there are 4 additional triangles, needed to
"close up" with the extra 4 triangles from Fig. 4 to get the eight triangles

FIGURE 3

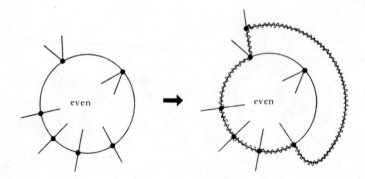

FIGURE 4

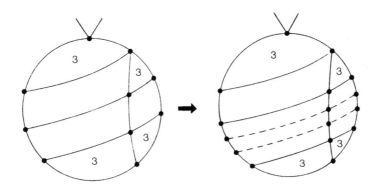

required in (2). The shells in Fig. 4 can be closed with one such as in Fig.
5, and are replicable with one 4-gon. Since this construction yields 4-
valent plane, 3-connected graphs, Steinitz's Theorem can be involved to complete the proof.

The number p_4^* obtained by this construction is not minimal for a given
solution of (2). The proper order of magnitude for a minimal value of p_4^*
is found in T. C. Enns' paper [2]. It is interesting to note that our proof
is in the spirit of Theorem 3, in that all values of p_4 beyond the one in the
construction are attainable. This is not the case for (1). For the solution
$p_3^* = 4$ of (1), only even p_6^* allow a polytope to be constructed.

If one writes down the Euler type relation for the valences t_i of a
tree, one obtains

FIGURE 5

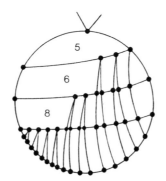

$$t_1 = 2 + \sum_{i \geq 2} (i - 2)t_i \tag{3}$$

The fact that the coefficient of t_2 vanishes suggests that one might be able to prove an analogue of Eberhard's Theorem for this case. This turns out to be the case. Details can be found in [10].

REFERENCES

1. V. Eberhard, *Zur Morphologie der Polyeder*, Teubner, Leipzig, 1891.

2. T. Enns, Convex 4-valent polytopes, *Discrete Math.* 30(1980), 227-234.

3. J. C. Fisher, An existence theorem for simple convex polyhedra, *Discrete Math.* 7(1974), 75-87.

4. B. Grünbaum, *Convex Polytopes*, Interscience, London, 1976.

5. B. Grünbaum, Some analogues of Eberhard's Theorem on convex polytopes, *Israel J. Math.* 6(1968), 398-411.

6. B. Grünbaum and J. Zaks, On the existence of certain planar maps, *Discrete Math.* 10(1974), 93-115.

7. S. Jendrol, A new proof of Eberhard's Theorem, *Acta Fac. Univ. Comescience Math.* 31(1975), 1-9.

8. E. Jucovic, On the face vector of a 4-valent, 3-polytope, *Stud. Sci. Math. Hung.* 8(1973), 53-57.

9. J. Malkevitch, Properties of Planar Graphs with Uniform Vertex and Face Structure, Ph.D. Thesis, University of Wisconsin, Madison, Wisconsin, 1969.

10. J. Malkevitch, Spanning trees in polytopal graphs, *Ann. N. Y. Academy of Science* 319(1979), 361-367.

11. J. Malkevitch, 3-valent, 3-polytopes with faces having fewer than 7 edges, *Ann. N. Y. Academy of Science* 175(1970), 285-286.

GRAPHICAL DIFFERENCE SETS AND PROJECTIVE PLANES*

Andrew Sobczyk

Department of Mathematical Sciences
Clemson University
Clemson, South Carolina

1. INTRODUCTION

One justification for presentation of a paper in combinatorics at a conference on convexity is of course the combinatorial structure of convex polytopes. In space R^d of dimension $d = 6k$ or $d = 6k + 2$, the skeleton-edge structure of a simplex is a Steiner system on $6k + 1$ or $6k + 3$ vertices; i.e.,
there are sets of edge-disjoint 2-faces or triangles which cover all of the
edges, each exactly once. At the problem sessions of this conference, the
author will pose questions like the following: Associate K_{15}, the complete
graph on fifteen vertices, with a regular planar polygon (see Fig. 1). The
edges are classified, by numbers of vertices "skipped," as 0, 1, 2, 3, 4, 5,
$6 \equiv 7$. All skip-2 edges are covered by the three equiskip-2 5-circuits,
and the remaining skips by two rotational series of fifteen triangles each,
the skip 0, 3, 4 triangles and the skip 1, 5, 6 triangles. Is there a 13-
polytope having fifteen vertices, which correspondingly has thirty triangular
2-faces and three pentagonal 2-faces, all edge-disjoint? (The edges of K_{15}
are uniquely covered also by the three 5-circuits and by fifteen quadruples
with outer skips 0, 3, 1, $7 \equiv 6$, inner skips 4, 5. In other terms, the

---

*Presented by title

FIGURE 1

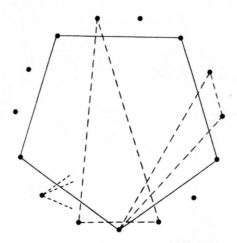

nonskip-2 edges are covered by the fifteen skip 0, 3, 4 triangles, and the fifteen 1, 5, 6 "forks" or basic matroid elements of K_{15}.)

Here is another question of a similar nature: The edges of K_{39} are
uniquely covered by two series of thirty-nine quadruples, those with respective exterior skips 0, 17, 8, 10; 1, 13, 16, 5, by thirteen equiskip-12 triangles, by the thirty-nine skip 2, 3, 6 triangles, and by the thirty-nine
skip 4, 9, 14 triangles. Is there a polytope, in R_d of appropriate maximal
possible dimension d, which correspondingly has seventy-eight rectangular 2-
faces and ninety-one triangular 2-faces? Perhaps not all of the triples and
quadruples can correspond to outer 2-faces; if so, this would reduce the
maximum possible dimension d. Skips for the several series may be clockwise
(c.) or counterclockwise (cc.); is it possible that there are polytopes for
some combinations (c., cc., cc., c., etc.) but not for others?

Also, the author believes that a viable topic is convex sets in the
vector spaces over the Galois fields GF(p) and $GF(p^k)$ [p prime, k an integer
> 1]. The points of the affine plane $AG(2,p^k)$ are the ordered pairs (x,y),
x, y $\in GF(p^k)$ [k $\geq$ 1]. As usual, the lines have equations of the form y =
mx + b, or x = c in case of infinite slope, m, b, c $\in GF(p^k)$. Thus if n = p^k,
$AG(2,n)$ has n^2 points, and the $n^2 + n$ lines fall into (n + 1) bundles of n
parallel lines each. For each point P(x,y), there is a pencil of (n + 1)
lines which pass through P. As soon as the author establishes the existence

of finite vector spaces R_d which are not the sets of d-tuples $(X_1, \ldots, X_d)$, $X_i \in GF(p^k)$ but which otherwise have the usual geometric properties of affine R_d (i.e., of finite affine spaces of orders different than p^k), the "convex-setters" (axiomatic and nonaxiomatic) may go to work in such spaces!

In view of results quoted, e.g., on pp. 27-28 of [2], with the usual axioms for affine and projective spaces, for $d \neq 2$ there are no affine or pro-jective d-spaces other than Galois (= Desarguesian). An (unusual) affine d-space, $d \neq 2$, such as the author visualizes, might involve matroids [6], nearfields, ternary rings, neofields [5], or other generalizations of Galois fields.

To extend $AG(2,n)$ to be the projective plane $PG(2,n)$, as usual for each slope there is adjoined a point at infinity; the line at infinity consists of all the points at infinity. Thus each line of $PG(2,n)$ has $(n + 1)$ points, and the total number of points, as well as the total number of lines, becomes $(n^2 + n + 1)$. For n other than $n = p^k$, the existence of $AG(2,n)$ is equiva-lent to the existence of a corresponding $PG(2,n)$.

2. ROTATIONAL MODELS FOR PROJECTIVE PLANES

A line contains all of its segments; in that respect a line having $(n+1)$ points is analogous to an n-simplex. The projective plane of order 2 may be viewed as the rotational Steiner system on seven vertices. The edges of K_7 are uniquely covered by the seven skip 0, 1, 2 triangles, which are the seven lines of $PG(2,2)$. The six lines of $AG(2,2)$ may be regarded as the six edges of K_4.

A rotational model for $PG(2,3)$ is provided by K_{13} associated with a polygon having thirteen vertices, the lines being the thirteen quadruples which have outer skips 1, 0, 3, 5, inner skips 2, 4. The quadruples uniquely cover all edges of K_{13}.

There are similar rotational models for all the projective planes $PG(2,n)$ over $GF(p^k)$ $[n = p^k]$, at least for $n = 4$, 5, 7, 8, 9, 11. For $n = 4$, the twenty-one lines are the twenty-one quintuples of K_{21} having outer skips 2, 0, 4, 1, 9, inner skips 5, 8, 6, 7, 3. For $n = 5$, the thirty-one lines (in K_{31}) are the thirty-one sextuples having outer skips 3, 5, 12, 0, 1, 4, inner skips 8, 9, 18 ≡ 11, 13, 2, 6, 7, 10, 15 ≡ 14. For $n = 7$, the outer skips are 4, 0, 1, 9, 18, 3, 6, 8, and inner skips the remaining skips of the skips 0 through 27 ≡ 28 of K_{57}; for $n = 8$, the outer skips in K_{73} are 0, 1, 3, 7, 15,

4, 17, 8, 9; for n = 9, in K_{91} they are 0, 1, 5, 17, 21, 6, 4, 15, 3, 9; for
n = 11, in K_{133} they are 6, 0, 1, 13, 11, 31, 18, 5, 4, 3, 17, 12.

3. ROTATIONAL MODELS FOR AFFINE PLANES

The Steiner system on K_9, which is AG(2,3), may be presented as the
eight skip 0, 1, 2 triangles of K_8, together with the four triangles formed
by joining an exterior vertex α (say α is the origin of AG(2,3)) with the
four skip-3 (diagonal) edges of K_8. If the vertices of K_8 are numbered cyclically as 1 through 8, the eight skip 0, 1, 2 triangles are 124, 235, 346,
457, 568, 671, 782, 813. It is easy to verify that the Steiner system is
AG(2,3): e.g., lines 124, 568, α37 are a typical bundle of three parallel
lines; the triangles which include α as a vertex are the pencil of four lines
through α; and through each point of K_8 there is a pencil of four lines, e.g.,
for vertex 1, the lines 124, 671, 813, α15.

Following is a different model for AG(2,3). Present the Steiner system
on K_9 as the join of an exterior triangle (line of AG(2,3)) and K_6. The equi-
skip-1 triangles of K_6 are the two parallel lines to the exterior line. The
three skip-2 diagonals of K_6, together with two disjoint sets of three of the
skip-0 edges, are three sets of three disjoint edges; the remaining nine lines
are the joins of each of the vertices of the exterior triangle to the edges
of one of the disjoint sets.

The affine planes AG(2,4) and AG(2,5) have models with exterior point
α, like the first given above for AG(2,3). For AG(2,4), five of the twenty
lines are the joins of exterior vertex α to the five equiskip-4 triangles of
K_{15}. The remaining fifteen lines are the fifteen quadruples which have exterior skips 3, 1, 0, 7 ≡ 6, inner skips 2, 5. For AG(2,5), α joined to the
six equiskip-5 quadruples of K_{24} forms a pencil of six lines through α. The
remaining twenty-four lines are the twenty-four quintuples of K_{24} with exterior skips 0, 2, 4, 1, 10, inner skips 3, 8, 9, 7, 6. (Skips 5, 11 are
covered by the equiskip-5 quadruples.)

The affine plane AG(2,5), like AG(2,3), also has a model formed by joining an exterior line of 5 points to K_{20}. The four parallel lines to the exterior line are the four equiskip-3 quintuples of K_{20}; they cover edges 7,
11. Five more lines are the five equiskip-4 quadruples with their points of
intersection with the exterior line. The remaining twenty lines are the
twenty quadruples of K_{20} which have outer skips (10 ≡ 8), 1, 0, 5, inner
skips 2, 6, joined to their points of intersection with the exterior line.
Here is a detailed verification by enumeration: Denote the points of the

exterior line as a, b, c, d, e, and the vertices of K_{20} (say in clockwise order) as 1,...,20. The five bundles of five parallel lines, besides the bundle first described, are the five columns of five lines, listed below.

16(11)(16)a	134(10)b	245(11)c
27(12)(17)b	578(14)a	689(15)b
38(13)(18)c	9(11)(12)(18)e	(10)(12)(13)(19)a
49(14)(19)d	(13)(15)(16)2d	(14)(16)(17)3e
5(10)(15)(20)e	(17)(19)(20)6c	(18)(20)17d

356(12)d	467(13)e
79(10)(16)c	8(10)(11)(17)d
(11)(13)(14)(20)b	(12)(14)(15)1c
(15)(17)(18)4a	(16)(18)(19)5b
(19)128e	(20)239a

4. THE AFFINE PLANE AG(2,7)

Regard K_{49} as the join of K_7 and K_{42}. Then K_7 and the six equiskip-5 heptagons of K_{42} are one bundle of seven parallel lines. The seven equiskip-6 hexagons determine one bundle of seven parallel lines which intersect the exterior line, K_7, in its seven points. Together the equiskip heptagons and hexagons cover the edges of skips 5, 6, 11, 13, 17, 20 of K_{42}. The remaining forty-two lines, which also intersect K_7, are determined by the forty-two hexagons which have the remaining skips 0, 1, 2, 3, 4, 7, 8, 9, 10, 12, 14, 15, 16, 18, 19 of K_{42}, as marked in Fig. 2.

FIGURE 2

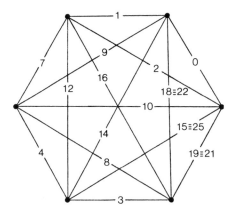

5. AFFINE PLANES OF NONPRIME-POWER ORDER

Until he discovered a slight arithmetical error, the author thought he had a model for AG(2,10), analogous to the one above for AG(2,7). That is,

K_{100} is regarded as the join of K_{10} and K_{90}. The nine equiskip-8 decagons of K_{90} and K_{10} are one bundle of ten parallel lines. The ten equiskip-9 nonagons of K_{90} determine another bundle of parallel lines which intersect the outer line, K_{10}, in its ten points. The equiskip decagons and nonagons cover the edges of skips 8, 9, 17, 19, 26, 29, 35, 39, 44. The remaining thirty-six skips of K_{90} should be covered by ninety rotational nonagons. On discovering his error, the author in haste assumed that if he could find twelve series of rotational triangles to cover the skips, the triangles could be assembled into nonagons as needed, since there is a Steiner system on K_q. He did indeed find a profusion of solutions for twelve triples (closed triangles) of skips covering the thirty-six skips, but so far has not succeeded in assembling one of the solutions for the triples into a nonagon. Computer help to decide this, and also to decide whether any of many other models for AG(2,10), AG(2,12), AG(2,15) proposed by the author are successful, has been promised by some of his colleagues who are good computer programmers.

By the Bruck-Ryser theorem, there does not exist a finite plane of order six. The edges of K_{43} are covered uniquely by the seven rotational series of triangles with skips 0, 16, 17; 1, 10, 12; 2, 15, 18; 3, 4, 8; 5, 13, 19; 6, 7, 14; 9, 11, 20. These cannot be assembled into a heptagon, for if they could be, we would have a model as in section 2 for PG(2,6). On the other hand, the nonagons in the model of PG(2,8) are dissectible into twelve rotational series of triangles which uniquely cover the edges of K_{73}. The triangles of course may be reassembled to form the nonagons.

Marshall Hall has proved that there does not exist a cyclic or rotational projective plane of order ten. The existence of a rotational affine plane with one exterior point in the lower order cases, in which cyclic projective planes exist, suggests to the author that Hall's result may imply also that there does not exist a rotational model with an exterior point for an affine plane of order ten. (Also, see [4].) But perhaps there is, as indicated above, a rotational affine model with an exterior line, exterior pair of lines, exterior pencil, or et cetera.

6. DUAL PLANES

The dual AG(2,n)* of AG(2,n) is PG(2,n) with deletion of one pencil and its "focus," but not of the other points of the lines of the pencil. In AG(2,n)*, not every two points are on a line, but every two lines are on a point. The number of lines per pencil is n (the same as the number of lines per bundle in AG(2,n)); the number of points on each line is (n + 1) (the same as the number of lines per pencil in AG(2,n)). For example, the obvious model for AG(2,3)* is the join $K_4 + K_8$: the nine lines are K_4, and the eight skip 0, 1, 2 triangles of K_8 with their points of intersection in K_4. Explicitly, if 1, 2, 3, 4 are the vertices of K_4, and 5, ..., 9, 0, α, β the vertices of K_8, then the nine lines of AG(2,3)* are 1234, 5681, 6792, 7803, 89α4, 90β1, 0α52, αβ63, β574. The twelve uncovered edges (pairs of points which are not on lines) are the diagonals 59, 60, 7α, 8β of K_8, and the eight bipartite edges 17, 28, 39, 40, 1α, 2β, 35, 46. The model for AG(2,4)* is the join $K_5 + K_{15}$. The sixteen lines are K_5 and the fifteen quadruples in K_{15} with outer skips 1, 2, 3, 5, inner skips 4, 6, together with their points of intersection in K_5. The pairs of points which are not on lines include the ends of each of the skip-0 edges.

For AG(2,5)* = $K_6 + K_{24}$, the twenty-five lines are K_6 and the twenty-four pentagons having outer skips 0, 1, 5, 10, 3, inner skips 2, 7, 6, 8, 4, with their points of intersection in K_6. The uncovered edges of K_{30} include the twenty-four skip-9 and twelve skip-11 edges in K_{24}

For the above model of AG(2,5)*, the pentagons with outer skips 1, 5, 3, 8, 2, inner skips 4, 7, 9, 10, 11, cannot serve as lines, because there are only twelve skip-11 or diagonal edges in K_{24}, so that the pairs of endpoints of the diagonals each would be on two lines. These latter pentagons, however, form some kind of partial combinatorial design on twenty-four objects: the pairs corresponding to the skip-0 and skip-6 edges are not covered; the pairs corresponding to skips-1, ..., 5, 7, ..., 10 are uniquely covered; and the pairs corresponding to the diagonals are each covered twice; by the twenty-four pentagons or blocks of five objects.

REFERENCES

1. A. A. Albert and R. Sandler, *An Introduction to Finite Projective Planes*, Holt, Rinehart and Winston, New York, 1968.

2. P. Dembowski, *Finite Geometries*, Springer-Verlag, New York, 1968.

3. J. W. P. Hirschfeld, *Projective Geometries Over Finite Fields*, Clarendon Press, Oxford, 1979.

4. A. J. Hoffman, Cyclic Affine Planes, *Can. J. of Math.* 4(1952), 295-301.

5. D. Frank Hsu, *Cyclic Neofields and Combinatorial Designs*, Springer-Verlag, New York, 1980.

6. D. J. A. Welsh, *Matroid Theory*, Academic Press, New York, 1976.

NOTES ADDED IN PROOF

Following is an easy matricial construction for AG(2,p), p prime, which shows the existence of a (graphical) model which is cyclic in a weaker sense than that of [4] and which does not require that there be an exterior point, or fixed point of a cyclic collineation. See Remark at p. 300 of [4]. In contrast with those in [4], the difference sets involved here are of a partial or mixed nature. (The author poses as a problem to find a similar matricial construction for the Galois $AG(2,p^k)$ $k > 1$.)

For the graphical-cyclical model for AG(2,3), numerate the points of K_9 as in Fig. 3. The vertical bundle of three parallel lines consists of the three equiskip-2 triangles, i.e., the rows of

$$\begin{pmatrix} 0 & 1 & 2 \\ 10 & 11 & 12 \\ 20 & 21 & 22 \end{pmatrix}$$

The horizontal bundle I consists of the lines

$$\begin{pmatrix} 0 & 10 & 20 \\ 1 & 11 & 21 \\ 2 & 12 & 22 \end{pmatrix}$$

The next bundle II,

$$\begin{pmatrix} 0 & 11 & 22 \\ 1 & 12 & 20 \\ 2 & 10 & 21 \end{pmatrix}$$

is obtained from I by taking the diagonal of I as first line, and counting cyclically to obtain two further lines.

The last bundle III,

$$\begin{pmatrix} 0 & 12 & 21 \\ 1 & 10 & 22 \\ 2 & 11 & 20 \end{pmatrix}$$

is obtained similarly from II. Referring to K_9, the lines of bundle I are three skip 0, 0, 1 triangles; the lines of bundle II are three skip 0, 3, 3 triangles; finally, the lines of III are three skip 1, 1, 3 triangles. The twelve triangles, in cycles of three, cover all of the skips 0, 1, 2, 3 ≡ 4 of K_9.

FIGURE 3

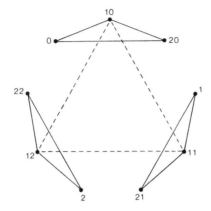

To further convey the idea, which produces a quasicyclic model for each prime p, consider AG(2,5). Disposing of excess notation and punctuation, the horizontal bundle I is

$$\begin{pmatrix} 0 & 0 & 0 & 0 & 0 \\ 1 & 1 & 1 & 1 & 1 \\ 2 & 2 & 2 & 2 & 2 \\ 3 & 3 & 3 & 3 & 3 \\ 4 & 4 & 4 & 4 & 4 \end{pmatrix}$$

with II, III, IV, and V, respectively,

$$\begin{pmatrix} 0 & 1 & 2 & 3 & 4 \\ 1 & 2 & 3 & 4 & 0 \\ 2 & 3 & 4 & 0 & 1 \\ 3 & 4 & 0 & 1 & 2 \\ 4 & 0 & 1 & 2 & 3 \end{pmatrix} \qquad \begin{pmatrix} 0 & 2 & 4 & 1 & 3 \\ 1 & 3 & 0 & 2 & 4 \\ 2 & 4 & 1 & 3 & 0 \\ 3 & 0 & 2 & 4 & 1 \\ 4 & 1 & 3 & 0 & 2 \end{pmatrix}$$

$$\begin{pmatrix} 0 & 3 & 1 & 4 & 2 \\ 1 & 4 & 2 & 0 & 3 \\ 2 & 0 & 3 & 1 & 4 \\ 3 & 1 & 4 & 2 & 0 \\ 4 & 2 & 0 & 3 & 1 \end{pmatrix} \qquad \begin{pmatrix} 0 & 4 & 3 & 2 & 1 \\ 1 & 0 & 4 & 3 & 2 \\ 2 & 1 & 0 & 4 & 3 \\ 3 & 2 & 1 & 0 & 4 \\ 4 & 3 & 2 & 1 & 0 \end{pmatrix}$$

In K_{25}, the vertical bundle consists of the five equiskip-4 pentagons, which cover all edges of skips 4, 9. Bundle I consists of five 0, 0, 0, 0, 3 pentagons, which cover edges of skips 0, 1, 2, 3. Bundle II consists of five 0, 5, 5, 5, 5, pentagons, which cover edges of skips 0, 5, 6, 11. Bundle III consists of five 2, 2, 7, 2, 7 pentagons, which cover edges of skips 2, 5, 7, 10. Bundle IV consists of five 6, 1, 6, 6, 1 pentagons, which cover edges of skips 1, 6, 8, 10. Bundle V consists of five 8, 3, 3, 3, 3, pentagons, which cover edges of skips 3, 7, 8, 11. Altogether, the thirty pentagons, in six cycles of five, cover all of the skips 0, 1, 2, 3, 4, 5, 6, 7, 8, 9, 10, 11 (twenty-five edges of each skip) of K_{25}.

For a start on an AG(2,10), the five equiskip-4 decogons (= complete subgraphs K_{10}) of K_{50} may be taken as five vertical lines. These use the edge of skips 4, 9, 14, 19, 24. A system of one hundred pentagons (= complete subgraphs K_5), formed from the twenty remaining skips of 0 through 23, perhaps in twenty cycles of five pentagons each, would be equivalent to three mutually orthogonal Latin squares of side 10. Whether there are three such squares is an open research question. Mostly by experimental search (unaided by computer), the author has found systems of one hundred edge-disjoint K_5's, but unfortunately for none found so far is the remaining subgraph of K_{50} the union (edgewise) of five K_{10}'s. Neither has the author been able to muster sufficient number-theoretic resources for a proof of nonexistence (say in cycles of five), nor has the earlier-promised computer programming help from his colleagues been forthcoming.

A system of one hundred pentagons, in cycles of ten, covering the edges of the twenty skips, is equivalent to four mutually orthogonal Latin squares. The author has produced systems of quadruples of K_{40}, in cycles of five, but which give rise only to two orthogonal squares; cycles of ten would imply ten bundles of parallel 4-point segments and therefore three mutually orthogonal squares.

TILING THE PLANE WITH INCONGRUENT REGULAR POLYGONS*

Hans Herda†

*Department of Applied Mathematics
The Weizman Institute of Science
Rehovot, Israel*

Professor Michael Edelstein asked this author how to tile the Euclidean plane with squares of integer side lengths all of which are incongruent. The question can be answered in a way which involves a perfect square and a geometric application of the Fibonacci numbers.

A perfect square is a square of integer side length which is tiled with more than one (but finitely many) component squares of integer side lengths all of which are incongruent. For more information, see the survey articles [3 and 5]. A perfect square is simple if it contains no proper subrectangle formed from more than one component square; otherwise, it is compound. It is known ([3], p. 884) that a compound perfect square must have at least 22 components. Duijvestijn's simple perfect square [2] (see Fig. 1) thus has the least possible number of components (21).

---

*Presented by title.

†*Current affiliation:* Department of Mathematics, Boston State College, Boston, Massachusetts

FIGURE 1.

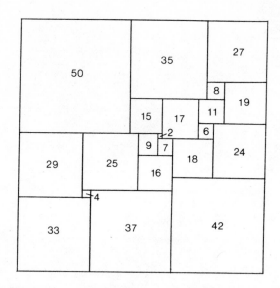

The Fibonacci numbers are defined recursively by $f_1 = 1$, $f_2 = 1$, and

$$f_{n+2} = f_n + f_{n+1} \qquad (n \geq 1) \qquad\qquad\qquad (*)$$

They are used in connection with the tiling shown in Fig. 2. Its nucleus is a 21-component Duijvestijn square, indicated by shading in Fig. 2, having side length $s = f_1 \cdot s = 112$, as in Fig. 1.

FIGURE 2.

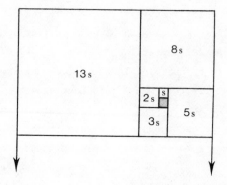

On top of this square we tile a one-component square s of side length $f_2 \cdot s = s = 112$, forming an overall rectangle of dimensions $2s$ by s. On the left side of this rectangle (the longer edge) we tile a square $2s$ of side length $f_3 \cdot s = 2s = 224$, forming an overall rectangle of dimensions $3s$ by $2s$. We now proceed counterclockwise as shown, each time tiling a

square f_ns onto the required longer edge of the last overall rectangle of
dimensions f_ns by f_{n-1}s, forming a new overall rectangle of dimensions f_{n+1}s
by f_ns (this follows from (*)). The tiling can continue indefinitely in this
way at each stage because f_ns = f_{n-1}s + f_{n-4}s + f_{n-3}s (this is used for n ≥ 5
and also follows from (*)). A closely related Fibonacci tiling for a single
quadrant of the plane (but beginning with two congruent squares) occurs in
([1], p. 305, Fig. 3).

    If we consider the center of the nuclear hatched square as the origin 0
of the plane, it is clear that the tiling eventually covers an arbitrary disc
centered at 0 and thus covers the whole plane. Finally, note that all the
component squares used in the tiling have integer side lengths and are incongruent.

    The tiling described above may be called static, since the tiles remain
fixed where placed, and the outward growth occurs at the periphery. It is
also interesting to consider a dynamic tiling. Start with a Duijvestijn
square. Its smallest component has side length 2. Enlarge it by a factor
of 56. The smallest component in the resulting square has side length 112.
Replace it by a Duijvestijn square. Now enlarge the whole configuration
again by a factor of 56. Repeat this process indefinitely, thus obtaining
the tiling. Here no tile remains fixed, outward growth occurs everywhere,
and it is impossible to write down a sequence of side lengths of squares used
in the tiling.

    The three-dimensional version of this tiling problem (due to D. F. Daykin)
is still unsolved: can 3-space be filled with cubes, all with integer side
lengths, no two cubes being the same size? ([4], p. 11).

    The plane can also be tiled with incongruent regular triangles and a
single regular hexagon, all having integer side lengths.

FIGURE 3.

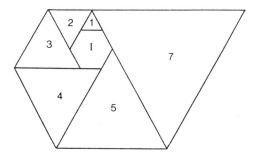

Begin with regular hexagon I (see Fig. 3) and tile regular triangles with side lengths 1, 2, 3, 4, 5 (and so named) counterclockwise around it as shown. Now tile a regular triangle with side length 7 along the sixth side of the hexagon. This counterclockwise tiling can be continued indefinitely to cover the plane. The recursion formula for the side lengths of the triangles is:

$$s_i = i \quad (i = 1 - 5), \quad s_6 = 7, \quad s_i = s_{i-1} + s_{i-5} \quad (i \geq 7)$$

ACKNOWLEDGMENT

The author would like to thank the Weizmann Institute of Science in Rehovot, Israel, where this work was done.

REFERENCES

1. A. Brousseau. Fibonacci numbers and geometry, *The Fibonacci Quarterly* 10(1972), 303-318,323.

2. A. J. W. Duijvestijn. Simple perfect squared square of lowest order, *J. of Combinatorial Theory* (B) 25(1978), 555-558.

3. N. D. Kazarinoff and R. Weitzenkamp. Squaring rectangles and squares, *Am. Math. Monthly* 80(1973), 877-888.

4. *Problems in Discrete Geometry*, collected and edited by William Moser with the help of participants of Discrete Geometry Week (July 1977, Oberwolfach) and other correspondents; 3rd ed., 1978, unpublished.

5. W. T. Tutte. The quest of the perfect square, *Am. Math. Monthly*, 72 (1965), No. 2, Part II, 29-35.

PROBLEMS*

1. (Steven Lay) Let K be a convex subset of E^n and suppose that the
interior of K is nonempty. Find necessary and sufficient conditions on K
for there to exist a nonconvex set S in E^n such that the kernel of S is equal
to K. In particular, does there exist a nonconvex subset of E^2 whose kernel
is the closed unit ball? (NOTE: Victor Klee has pointed out that this problem was originally posed by Fejes-Toth and was essentially solved by Klee
in his paper, A theorem on convex kernels, *Mathematika* 12(1965), 89-93. His
result applies to the case when K is a closed, convex set in a separable Ba-
nach space. Marilyn Breen independently obtained essentially the same result
for R^d and also a result on nonclosed sets in her paper "Admissable Kernels
for Starshaped Sets," *Proc. of AMS*, 82, No. 4 (1981), 622-628. David Kay has

_____

Editorial Note: The first two problems stated above were the only ones
presented at the Problem Session of the conference, the majority of the problems having been posed during the presentation of papers. Those problems
appear in previous chapters of these *Proceedings* and will not be duplicated
here. The other problems are the result of another convexity conference held
only one month later (Special Session on Combinatorial Geometry and Convex
sets, 777th Meeting of the American Mathematical Society, Davis, California,
April, 1980). The organizers of that conference, G. D. Chakerian and David
Barnette, and the editors of these *Proceedings* have agreed to include those
problems here as a service to the mathematical community.

recently generalized Klee's result for noncomplete spaces in "Starshaped Sets with Prescribed Convex Kernels in Separable, Normed Linear Spaces", submitted for publication. The original problem remains for sets in the general setting of linear topological spaces, and almost no results of a general nature have been obtained for the case when K is closed, even for E^n).

2. (David Kay) Define a *convex representation* of an alignment of X to be a one-to-one convexity-preserving map from X to the relativized convexity space of a subset of R^d for some d using the convex hull in R^d. A representation is *affine* if we use the affine hull in R^d. Two problems, concerning which almost nothing is known, are:

   (a) Characterize those alignments which have a convex representation.

   (b) Characterize those affine alignments (i.e., matroids) which have an affine representation.

3. (Richard Gardner) Suppose K is a compact convex subset of the plane, and three lines L_1, L_2, L_3 meet at a point in K, dividing K into 6 regions, labelled cyclically by $(x_1,y_2,x_3,y_1,x_2,y_3)$. If $|E|$ denotes the area of E, it may be shown that

$$\frac{|x_1|}{|y_1|} + \frac{|x_2|}{|y_2|} + \frac{|x_3|}{|y_3|} \geq \frac{3}{2} \tag{1}$$

$$\frac{|x_2 + x_3|}{|y_1|} + \frac{|x_1 + x_3|}{|y_2|} + \frac{|x_1 + x_2|}{|y_3|} \geq 3 \tag{2}$$

(It suffices to take K to be a triangle.) It is conjectured that equality in (1) and (2) above holds if and only if K is a triangle and the lines L_1, L_2, L_3 pass through the centroid of K, parallel to the sides of K. (See the article R. J. Gardner, S. Kwapien and D. Laurie, Some inequalities related to compact convex sets, to be published.)

4. (Jacob E. Goodman) This problem was conveyed to me by R. Pollack, who heard it from A. Lax, who originally got it from a draftsman: If three rays meet at a point, determine all the triangles having vertices on those three rays.

If the rays are mutually perpendicular, it is not hard to see that the answer is: all acute triangles. D. Kay (private communication) has generalized this to the following partial solution: If the rays make pairwise angles ρ, σ, τ, each $\geqslant \pi/2$, then the answer is: all triangles whose angles α, β, γ satisfy $\max(\alpha,\beta,\gamma) < \max(\rho,\sigma,\tau)$. I know of no complete solution.

5. (Jacob E. Goodman) Given a (non-convex) plane polygon Π, consider the problem of finding a convex subset K of largest area. It is not hard to see that K exists, is necessarily itself a polygon, and has area $\geqslant \dfrac{1}{n-2}$; moreover, there are n-gons Π for which K has area only arbitrarily little more than $\dfrac{1}{n-2}$. The question is: How does one find such a polygon K inside Π? There is a (finite) algorithm described in the paper cited below, in the case $n \leqslant 5$, for locating K (its sides cannot always be obtained by prolonging the sides of Π!). The problem then is to extend this to an algorithm which works if $n > 5$. (See J. E. Goodman, On the largest convex polygon contained in a nonconvex n-gon, or how to peel a potato, *Geometriae Dedicata*, to be published.)

6. (Branko Grünbaum) It is well known that, for each $n \geqslant 3$, the euclidean plane E^2 can be tiled by convex polygons each of which has n sides; thus each convex polygon is a *combinatorial prototile* of some tiling of the plane. In 1975 the following problem was communicated to me by Ludwig Danzer: Is every (bounded) convex polyhedron in E^3 a combinatorial prototile of some tiling in E^3?

7. (Branko Grünbaum) The situation in Problem 6 is drastically altered if instead of "combinatorial prototiles" we consider "congruence prototiles." If a convex polygon is a congruence prototile it has at most 6 sides, but not every convex polygon with at most six sides is a congruence prototile (e.g., the regular pentagon). No characterization is known for those that are (see D. Schattschneider, Tiling the plane with congruent pentagons, *Math. Mag.* 51 (1978), 29-44). Determine the largest number of faces (or of vertices) possible in a convex polyhedron which is a congruence prototile for E^3. (Note: The crystallographer Peter Engel from Bern recently found examples of such polyhedra with up to 38 faces and up to 70 vertices; see the account in B. Grünbaum and G. C. Shephard, Tilings with congruent tiles, *Bulletin of Amer. Math. Soc.* 3(1980), to be published.)

8.   (J. M. Wills)   Let K_1^d, K_2^d, K_3^d be the three convex bodies which define the three classical metrics: Euclidean (ball), cube and octahedron and with d dimensional volume $V(K_i^d) = 1$, $d = 2, 3, \ldots$ with ω_d the volume of the unit sphere. Thus,

$$K_1^d = \{(x_1,\ldots,x_d) / (\textstyle\sum_{i=1}^{d} x_i^2)^{1/2} \leqslant \omega_d^{-1/d}\}$$

$$K_2^d = \{(x_1,\ldots,x_d) / \textstyle\sum_{i=1}^{d} |x_i|^{1/2} \leqslant (d!)^{1/2}\}$$

$$K_3^d = \{(x_1,\ldots,x_d) / \max_{i=1,\ldots,d} |x_i| \leqslant 1/2\}$$

If $K_{ij}^d = K_i^d \cap K_j^d$, $i < j$, then trivially $0 < V(K_{ij}^d) < 1$.

(a)   Do the three limits $\lim\limits_{d\to\infty} V(K_{ij}^d)$ exist?

(b)   If yes, are they > 0 or $= 0$?

9.   (G. L. Alexanderson and John E. Wetzel)   Let f_k be the number of k-dimensional flats and $C'(r)$ the number of bounded r-dimensional cells formed by an arbitrary arrangement of hyperplanes in E^d, and consider the inequalities

$$C'(r) \geqslant \sum_{k=0}^{r} (-1)^k \binom{d - k}{k - r} f_k \qquad\qquad (*)$$

for $0 \leqslant r \leqslant d$. In our paper, "Arrangements of planes in space," to appear in *Discrete Mathematics*, we prove (*) for all non-degenerate arrangements, i.e., arrangements whose normals span E^d, in the cases $d = 2$ and $d = 3$; and we determine the arrangements for which the equalities hold.

Although the inequalities (*) seem to hold for a wide class of arrangements in E^4, they do not hold for all non-degenerate 4-arrangements. Precisely when are they true for $d = 4$, and what is the situation for $d > 4$?

10.   (Paul Erdös; posed jointly with Szemerédi)   Let $x_1,\ldots,x_n$ be n points in the unit square (or unit circle). Denote by $D(x_1,\ldots,x_n)$ the smallest distance $d(x_i,x_j)$, $1 \leqslant i < j \leqslant n$ and by $\alpha(x_1,\ldots,x_n)$ the smallest angle determined by these n points. We conjecture that if $D(x_1,\ldots,x_n) > \dfrac{\varepsilon}{\sqrt{n}}$, then $\alpha(x_1,\ldots,x_n) = o\left(\dfrac{1}{n}\right)$. More generally, put

$$F(n) = \max_{x_1,\ldots,x_n} (D(x_1,\ldots,x_n) \cdot \alpha(x_1,\ldots,x_n))$$

Prove $F(n) = o\left(\dfrac{1}{n^{3/2}}\right)$ and determine $F(n)$ as well as you can. $F(n) > cn^{-2}$ is obvious.

Background: Let $x_1 \cdot, , , , \cdot x_n$ be n points in the unit square (or circle). $A(x_1, \ldots, x_n)$ is the smallest area of all the $\binom{n}{3}$ triangles (x_i, x_j, x_ℓ). Put

$$A(n) = \max_{x_1, \ldots, x_n} A(x_1, \ldots, x_n).$$

The determination of $A(n)$ is known as *Heilbronn's problem*. Heilbronn claimed he only transmitted it, but since he is unfortunately cured of our incurable disease we cannot find out. It was conjectured that

$$\frac{c_1}{n^2} < A(n) < \frac{c_2}{n^2} \tag{1}$$

K. F. Roth first proved $A(n) = o\left(\dfrac{1}{n}\right)$. This was sharpened by W. Schmidt and later Roth. I observed $A(n) > c_1/n^2$. Recently Komlós, Pintz and Szemerédi proved

$$\frac{c_3 \log n}{n^2} < A(n) < \frac{c_4}{n^{8/7}}.$$

That is, (1) is *false*. The estimation of $F(n)$ may help with $A(n)$ but is of interest in itself.

11. (Paul Erdös; posed jointly with Corrádi and Hajnal) Let $x_1, \ldots, x_n$ be n points in the plane. If no three are on a line, then trivially they determine at least n - 2 distinct angles. Does this remain true if we only assume that not all the points are on a line? Is it true that they determine at least cn distinct angles?

12. (Paul Erdös) I conjectured and Altman proved that if $(x_1, \ldots, x_n)$ is a convex n-gon then there are at least $\left[\dfrac{n}{2}\right]$ distinct numbers among the $d(x_i, x_j)$. Szemerédi conjectured that the same result holds if no three of the x_i's are on a line. He only proved this with $\left[\dfrac{n}{3}\right]$ instead of $\left[\dfrac{n}{2}\right]$. (See P. Erdös, On some problems of elementary and combinatorial geometry, *Annali di Mat. ser. IV* 103(1975), 99-108.)

13. (Paul Erdös) Let $x_1, \ldots, x_n$ be n points in the plane, at most n − k on a line. Join any two of them. Is it true that they determine at least ckn distinct lines where c is an absolute constant? For $k < 3\sqrt{n}$ this and more follows from a result of L. M. Kelly and W. Moser. (See L. M. Kelly and W. Moser, On the number of ordinary lines determined by n points, *Canad. J. Math.* 10(1958), 210-219.)

14. (Paul Erdös) Let $x_1, \ldots, x_n$) be n points in the plane, no r + 1 on a line. Let $f_r(n)$ be the largest integer for which there are $f_r(n)$ lines which contain exactly r of the points. Sylvester proved that $f_3(n) = (1 + o(1))n^2/6$. His results were sharpened by Burr, Grünbaum and Sloane. I conjectured that for r > 3

$$\frac{f_r(n)}{n^2} \to 0$$

Grünbaum proved that $f_r(n) > cn^{1 + \frac{1}{r-2}}$, sharpening a previous result of Kárteszi which stated $f_r(n) > cn \log n$. (See S. A. Burr, B. Grünbaum and N. J. A. Sloane, The orchard problem, *Geometriae Dedicata* 2(1974), 397-424 and B. Brünbaum, New Views of old questions of combinatorial geometry, *Coll. Internazionale Teorie Combinatorie, Roma 1973, Accad. Naz. Lincei, Tomo I* (1976), 451-468.)

15. (Paul Erdös) Let $x_1, \ldots, x_n$ be n points in the plane not all on a line. Join every two of them. We then obtain the lines $L_1, \ldots, L_m$. Assume that L_i has α_i points, $\alpha_1 \geq \alpha_2 \geq \cdots \geq \alpha_m$. A classical theorem of Gallai states that $\alpha_m = 2$. Gallai's theorem easily implies m ≥ n. Denote by F(n) the number of possible choices of $\{\alpha_1, \ldots, \alpha_m\}$. I conjecture that

$$F(n) < e^{Cn^{1/2}}$$

It is not hard to see that apart from the value of C this--if true--is best possible. (For further problems of this kind see P. Erdös, Some combinatorial problems in geometry, *Lecture Notes in Math 792, Geometry and Differential Geometry, Proc., Haifa, Israel* (1979), 46-53; this paper contains extensive references.)

AUTHOR INDEX

PROBLEM INDEX

SUBJECT INDEX

A

Abstract convexity, 120, 164, 167
Affine plane, 216, 219
  matricial construction for, 222
  rotational model for, 218
Alignment (*also*, Aligned space),
  120, 167, 173, 174
  affine (in R^d), 123
  box, 130, 167, 176
  category of alignments, 131
  free, 128
  ordinary (of R^d), 123
Antimatroid, 124, 140, 144
  anti-exchange law of, 125
  variety of, 135
Arrangement(s):
  of lines, 73
  nonstretchable, 73-75
  of pseudolines, 77-79
Axiomatic convexity, 120, 164, 167

B

Barbier's Theorem, 59
Basis:
  positive, 3
  of a variety, 142
Billiards:
  bibliography for, 104
  convex caustic for, 86, 98
Blaschke-Lebesgue Theorem, 61
Blaschke's relation, 60
Box alignment, 130
  in R^d, 167, 176
Box, in R^n, 204
Breadth, 127
Bruck-Ryser, theorem of, 217

C

Carathéodory number:
  in abstract convexity, 167, 168,
    176
  in alignments, 122, 127
  equal to unity, 140
  and minimal forbidden subspaces,
    143, 144
  used to define varieties, 129, 137
Carathéodory's Theorem:
  in abstract convexity, 118
  in R^2, 77
Category of alignments, 131
Cap body, 89
Caustic curves, 86
Caustics, convex (*see also* Billiards):
  in R^2, 86-98
  in R^3, 99-103
Chinese Remainder Theorem, 117
Circular disk, characterization of, 63
Closure:
  algebraic closure system(s), 114,
    115, 174
  topological, 115, 116
Compact element:
  convex body, 57, 193
  element, 139
Complex:
  Cohen-Macaulay, 29
  2-cell, 7
Constant width, 58, 63
Contraction, 132
Contractible varieties, 132
Convex:
  body, 57
  caustic, for billiards, 86, 98
  compact body, 88, 193
  cone, generating, 194
  subsets, of convexity structure, 120
Copoint, of matroids, 116, 128

239